Collins

AQA GCSE 9-1

Physics

Physics

Revision guide

AQA
GCSE 9-1

Revision guide

Nathan Goodman

Contents

HT Higher Tier Content

Contents

HT Higher Tier Content

Contents

HT Higher Tier Content

Contents

HT Higher Tier Content

Recap of KS3 Key Concepts

Figure 1

Wall

Magnet fixed to truck

Magnet fixed to wall

Wooden truck | S N | | N S |

Table

1. In **Figure 1** the wooden truck is held so it does not move.

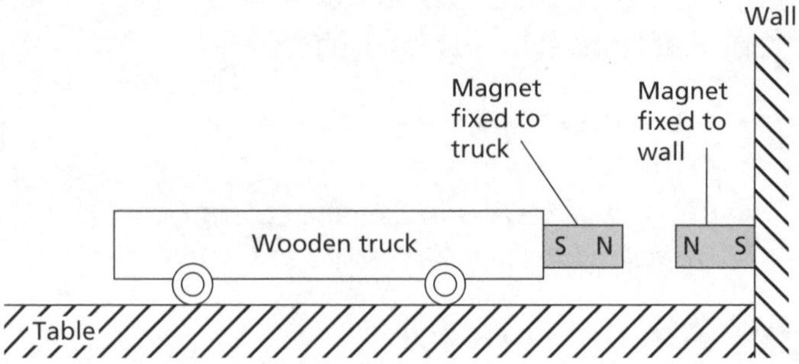

   a) Describe the motion of the truck when it is released. [1]

   b) What effect will friction have on the truck? [1]

2. During summer, people use small hand-held fans to keep themselves cool.

   a) Draw a series circuit containing a battery, switch and motor to show how one of these fans could be wired. [3]

   b) Which component provides energy for the circuit? [1]

   c) A bulb is added in series to the circuit.

   How will this affect the motor? [1]

3. A series circuit contains three bulbs.

   What will happen to the other bulbs if the bulb in the middle breaks?
   Tick **one** box.

   Both bulbs will go out. ☐

   Both bulbs will get brighter. ☐

   The one after the broken bulb will go out. ☐

   Both bulbs will stay the same. ☐ [1]

4. A parallel circuit contains three bulbs.

   What will happen to the other bulbs if the bulb in middle breaks?
   Tick **one** box.

   Both bulbs will go out. ☐

   Both bulbs will get brighter. ☐

   The one after the broken bulb will go out. ☐

   Both bulbs will stay the same. ☐ [1]

5 Oil is described as a non-renewable energy resource.

a) Explain what is meant by this. [2]

b) Give **two** other non-renewable energy resources. [2]

6 Choose words from the list to complete the paragraph.

chemical   electrical   gravitational   kinetic   light   sound   thermal

A bicycle light uses a generator powered by the turning of the wheels. As a cyclist pedals,

.... energy in her muscles is changed to kinetic energy.

When the generator turns, kinetic energy is changed to useful .... energy in the wires.

This energy in the wires is changed to useful .... energy by the bulb.

When the light is on, some of the energy in the bulb is wasted as .... energy. [4]

7 When a bird flies at a constant height, there is a downward force of 30N on the bird.

How large is the upwards force on the bird? [1]

8 **Figure 2** shows four forces acting on an aeroplane in flight.

Figure 2

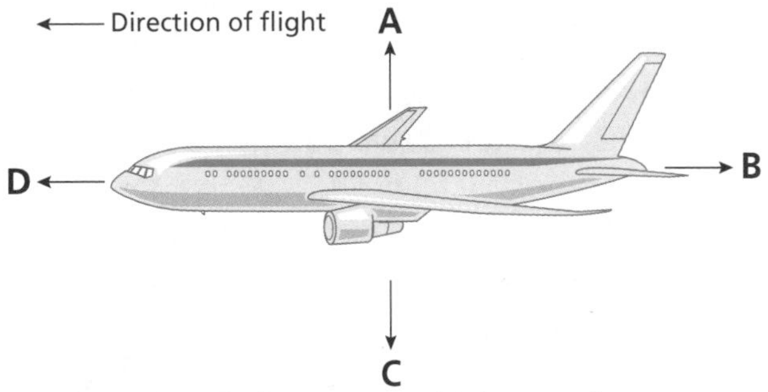

← Direction of flight   A

D ←   → B

C

a) Which arrow represents air resistance? [1]

b) When the aeroplane is descending, what can be said about forces **A** and **C**? [1]

c) When the plane is flying at a constant speed in the direction shown, which **two** forces must be balanced? [1]

d) Just before takeoff, the aeroplane is speeding up along the ground.

What must this mean about the size of forces **B** and **D**? [1]

9 Two cyclists are riding along a dark road at night.
One is wearing black clothes and the other is wearing light-coloured clothes.
A car approaches the cyclists with its headlights on.

a) What happens to the light when it reaches the light-coloured clothes? [1]

b) Explain why it is more difficult for the driver to see the cyclist in the black clothes. [3]

Total Marks .... / 26

Forces – An Introduction

You must be able to:

- Describe the difference between a vector and a scalar quantity
- Use vectors to describe the forces involved when objects interact
- Calculate the resultant of two forces that act in a straight line
- Explain the difference between mass and weight
- Recognise and use the symbol for proportionality
- **HT** Use vector diagrams to show resolution and addition of forces.

Scalar and Vector Quantities

- A scalar quantity has magnitude (size) only, e.g. number of apples.
- A vector quantity has magnitude and direction, e.g. velocity, which shows the speed *and* the direction of travel.
- Arrows can be used to represent vector quantities :
  - the length of the arrow shows the magnitude
  - the arrow points in the direction that the vector quantity is acting.
- Forces are vector quantities.
- The diagram shows the forces acting on a boat. The arrows indicate the direction they are acting.

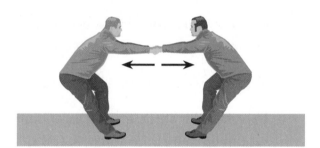

Water pushes boat up (buoyancy)

Resistive force

Driving force

Gravity pulls boat down (weight)

Contact and Non-Contact Forces

- A force occurs when two or more objects interact.
- Forces are either:
  - **contact forces** – the objects are actually touching, e.g. the tension as two people pull against one another
  - **non-contact forces** – the objects are not touching, e.g. the force of gravity acts even when the objects are not touching.

Contact Forces	Non-Contact Forces
Friction Air-resistance / drag Tension Normal contact force Upthrust	Gravitational force Electrostatic force Magnetic force

> **Key Point**
>
> A force is a vector quantity. It occurs when objects interact.

Gravity

- Gravity is a force of attraction between all masses.
- The force of gravity close to Earth is due to the gravitational field around the planet.
- The mass of an object is related to the amount of matter it contains and is constant.
- Weight is the force acting on an object due to gravity.
- The weight of an object depends on the gravitational field strength where the object is and is directly proportional to its mass.

- This symbol is used to indicate two things are proportional: ∝.

Revise

LEARN

weight = mass × gravitational field strength

$$W = mg$$

Weight (W) is measured in newtons (N).
Mass (m) is measured in kilograms (kg).
Gravitational field strength (g) is measured in newtons per kilogram (N/kg).

Resultant Forces

- When more than one force acts on an object, these forces can be seen as a single force that has the same effect as all the forces acting together.
- This is called the resultant force.

> **Key Point**
>
> The gravitational field strength (g) on Earth is 10N/kg, so a student with a mass of 50kg has a weight of (50 × 10 =) 500N.

HT Vector Diagrams

- A free body diagram can be used to show different forces acting on an object (see the diagram on the boat on page 8).
- Scale vector diagrams are used to illustrate the overall effect when more than one force acts on an object:
  - The forces are added together to find a single resultant force, including both magnitude and direction.
  - The vectors are added head to tail and a resultant force arrow is drawn.

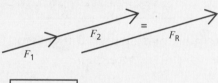

$$F_R = F_1 + F_2$$

- Scale vector diagrams can also be used when a force is acting in a diagonal direction:
  - Expressing the diagonal force as two forces at right-angles to each other can help to work out what effect the force will have.
  - The force F_R can be broken down into F_1 and F_2.
  - F_1 is the same length as the length of F_R in the horizontal direction.
  - F_2 is the same length as the length of F_R in the vertical direction.
  - F_R is also the vector found by adding F_1 and F_2 head to tail.

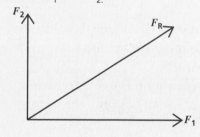

10N
Frictional forces

15N
Pushing force

Resultant force ⟶ 5N

> **Key Point**
>
> Weight is a force that can be measured using a newtonmeter (a calibrated spring-balance). The unit of measurement is newtons (N).
> weight ∝ mass

Key Words

scalar
vector
force
contact force
non-contact force
gravity
mass
weight
resultant
HT free body diagram

Quick Test

1. How is a scalar quantity different from a vector quantity?
2. Use an example to explain what is meant by 'non-contact force'.
3. An astronaut with a mass of 80kg stands on the moon. What is his weight? (g = 1.6N/kg)
4. A car travels in a straight line to the east with a driving force of 500N. If total frictional forces are 400N, what is the resultant force and in which direction does it act?

Forces in Action

You must be able to:

- Describe the energy transfers involved when work is done
- Explain why changing the shape of an object can only happen when more than one force is applied to the object
- Interpret data showing the relationship between force and extension
- Perform force calculations for balanced objects
- Explain how levers and gears work.

Work Done and Energy Transfer

- When a force causes an object to move, **work** is done on the object.
- This is because it requires energy to move the object.
- One joule of work is done when a force of one newton causes a displacement of one metre: 1 joule = 1 newton metre.

LEARN

work done = force × distance (moved along the line of action of the force)

$$W = Fs$$

> Work done (W) is measured in joules (J).
> Force (F) is measured in newtons (N).
> Distance (s) is measured in metres (m).

- When work is done, energy transfers take place within the system, e.g. work done to overcome friction causes an increase in heat energy.

> ### Key Point
>
> Overcoming forces requires energy. When a force is used to move an object, work is done on the object. The movement of the object is called displacement.

Forces and Elasticity

- To change the shape of an object, more than one force must be applied, e.g. a spring must be pulled from both ends to stretch it.
- If the object returns to its original shape after the forces are removed, it was **elastically deformed**.
- If the object does *not* return to its original shape, it has been **inelastically deformed**.
- The **extension** of an elastic object is directly proportional to the applied force, i.e. they have a linear relationship and produce a straight line on a force–extension graph.
- However, once the **limit of proportionality** has been exceeded:
  - doubling the force will no longer exactly double the extension
  - the relationship becomes non-linear
  - a force–extension graph will stop being a straight line.
- This equation applies to the linear section of a force–extension graph:

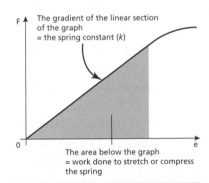

The gradient of the linear section of the graph = the spring constant (k)

The area below the graph = work done to stretch or compress the spring

LEARN

force = spring constant × extension

$$F = ke$$

> Force (F) is measured in newtons (N).
> Spring constant (k) is measured in newtons per metre (N/m).
> Extension (e) is measured in metres (m).

- This also applies to the **compression** of an elastic object.
- The **spring constant** indicates how easy it is to stretch or compress a spring – the higher the spring constant, the stiffer the spring.
- A force that stretches or compresses a spring stores elastic potential energy in the spring.
- The amount of work done and the energy stored are equal, provided the spring does not go past the limit of proportionality.

> ### Key Point
>
> The work done to stretch or compress a spring is equal to the energy stored in the spring, provided the spring has not been inelastically deformed.

Investigate the relationship between force and extension of a spring.

Sample Method	Considerations, Mistakes and Errors
1. Set up the equipment as shown. 2. Add 100g (1N) to the mass holder. 3. Measure the extension of the spring and record the result. 4. Repeat steps 2 to 3 for a range of masses from 1N to 10N.	• The extension is the total increase in length from the original unloaded length. It is *not* the total length or the increase each time. • Adding too many masses can stretch the spring too far, which means repeat measurements cannot be made.

Variables	Hazards and Risks
• The independent variable is the one deliberately changed – in this case, the force on the spring. • The dependent variable is the one that is measured – the extension.	• The biggest hazard in this experiment is masses falling onto the experimenter's feet. To minimise this risk, keep masses to the minimum needed for a good range of results.

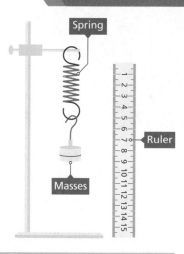

Spring

Ruler

Masses

Moments, Levers and Gears

- When a force causes an object to rotate about a **pivot** point, the turning effect is called **moment** of a force.

$$\text{moment of a force} = \text{force} \times \text{distance}$$

$$M = Fd$$

Moment of a force (M) is measured in newton metres (Nm).
Force (F) is measured in newtons (N).
Distance (d) is the perpendicular distance from the pivot to the line of action of the force in metres (m).

- If an object is balanced, the total clockwise moment about the pivot equals the total anticlockwise moment about that pivot:

$$F_1 \times d_1 = F_2 \times d_2$$

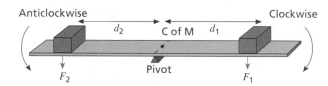

Anticlockwise d_2 C of M d_1 Clockwise

F_2 Pivot F_1

Perpendicular distance between the line of action of the force and the pivot

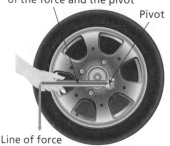

Pivot

Line of force

A force of 500N is positioned 1m from the middle of a seesaw. It is balanced by another force 2m from the middle. How large is the second force?

$$500 \times 1 = F_2 \times 2$$

Substitute the given values into the equation $F_1 \times d_1 = F_2 \times d_2$.

$$F_2 = \frac{500 \times 1}{2} = 250N$$

Divide both sides by 2 to leave F_2 on its own.

- **Levers** and **gears** can be used to:
  - transmit the rotational effects of forces
  - magnify either the size of the applied force or the distance the force moves over.
- When the applied force moves further than the transmitted force, the force is increased.
- When the applied force is bigger than the transmitted force, the distance is increased.

Key Point

The range in an experiment needs to be large enough to show a pattern.

Key Words

work
elastically deformed
inelastically deformed
extension
limit of proportionality
compression
spring constant
pivot
moment
lever
gear

Quick Test

1. Calculate the work done when a force of 200N is used to move an object 50cm.
2. A plank is pivoted in the middle. A 30N force is placed 50cm from the pivot. How far from the pivot, on the other side, would a 40N force need to be applied to balance the plank?

Pressure and Pressure Differences

You must be able to:

- Recall and use the equation for pressure
- Describe a simple model of the Earth's atmosphere
- Explain why atmospheric pressure varies with height
- **HT** Explain the effect of fluid density on pressure
- **HT** Calculate difference in pressure at different depths in a liquid
- **HT** Explain the origin of upthrust.

Pressure in a Fluid

- A **fluid** can either be a liquid or a gas.
- As particles move around in a fluid, they collide with the surface of objects in the fluid or the surface of the container.
- The collisions create a force **normal** (at right-angles) to the surface.
- The link between **pressure**, force and area is described with the following equation:

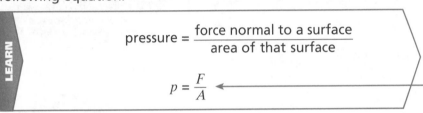

$$pressure = \frac{force\ normal\ to\ a\ surface}{area\ of\ that\ surface}$$

$$p = \frac{F}{A}$$

LEARN

- If the pressure acts on a bigger area, it will produce a larger force.

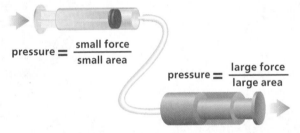

$$pressure = \frac{small\ force}{small\ area}$$

$$pressure = \frac{large\ force}{large\ area}$$

Key Point

Pressure is caused by particles colliding with a surface.

Pressure (*p*) is measured in pascals (Pa).
Force (*F*) is measured in newtons (N).
Area (*A*) is measured in metres squared (m²).

Atmospheric Pressure

- The **atmosphere** is a relatively thin layer of air around the Earth.
- The greater the **altitude**, the less **dense** the atmosphere and the lower the atmospheric pressure.
- At a high altitude there is less air above a surface than at lower altitudes, so there is a smaller weight of air acting on the surface and the equation $p = \frac{F}{A}$ will result in a lower pressure.

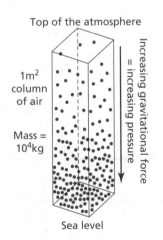

Top of the atmosphere

1m² column of air

Mass = 10⁴kg

Increasing gravitational force = increasing pressure

Sea level

HT Pressure in a Column of Liquid

- Pressure at a particular point in a column of liquid depends on:
  - the height of the column above the point
  - the density of the liquid.
- The higher the column and the more dense the liquid:
  - the greater the weight above the point
  - the greater the force on the surface at that point
  - the greater the pressure.

Key Point

The particle model explaining pressure is a very good example of a model explaining a scientific idea.

- The pressure can be calculated with the equation below:

LEARN **HT**

pressure = height of the column × density of the liquid × gravitational field strength

$$p = h\rho g$$

A diver descends from the surface to a depth of 25m. The gravitational field strength is 10N/kg and the density of water is 1000kg/m³.

Work out the increase in pressure.

$$p = h\rho g = 25 \times 1000 \times 10$$
$$p = 250\,000\text{Pa}$$

Substitute the given values into the equation.

Don't forget the units.

HT Upthrust

- When an object is submerged in a liquid, there is a greater height of liquid above the bottom surface than above the top surface.
- The bottom surface experiences a greater pressure than the top surface and this creates a resultant force upwards.
- The upwards force exerted by a fluid on a submerged object is called upthrust.
- An object floats when its weight is equal to the upthrust and sinks when its weight is greater than the upthrust.
- The density of an object indicates if it will float or sink.
- An object less dense than the liquid:
  - displaces a volume of liquid greater than its own weight so it will rise to the surface
  - will float with some of the object remaining below surface
  - displaces liquid of equal weight to the object.
- If an object has a low density, more of the object will remain above the surface.
- An object denser than the surrounding liquid cannot displace enough liquid to equal its own weight so it sinks.
- No matter how dense an object is, the size of the upthrust is equal to the weight of liquid displaced.

HT Key Point

The deeper an object is submerged the greater the pressure.

A more dense liquid exerts a greater pressure.

The Density of an Object Affects if it will Float or Sink

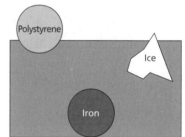

Quick Test

1. Describe how the particles in a fluid create pressure.
2. Atmospheric pressure is 100 000Pa. What force does this apply on an area of 0.1m²?
3. **HT** An object floats in water. 90% of the object is below the water line. The density of water is 1000kg/m³.
   Estimate the density of the object.
4. **HT** Find the pressure difference between a submarine at a depth of 100m and another submarine at a depth of 300m.
   density of water = 1000kg/m³,
   gravitational field strength = 10N/kg

Key Words

fluid
normal
pressure
atmosphere
altitude
dense
HT upthrust
HT displace

Forces and Motion

You must be able to:

- Describe displacement in terms of both magnitude and direction
- Recall typical values for common speeds
- Calculate speed from measurements of distance and time
- **HT** Give examples of objects with constant speed but changing velocity
- Explain the motion of objects using Newton's first law
- Draw and interpret distance–time graphs.

Distance and Displacement

- Distance is a scalar quantity:
  - It is how far an object moves.
  - It does not take into account the direction an object is travelling in or even if it ends up back where it started.
- Displacement is a vector quantity:
  - It has a magnitude, which describes how far the object has travelled from the origin, measured in a straight line.
  - It has a direction, which is the direction of the straight line.

Path of travel

Displacement

> **Key Point**
>
> Distance and speed are scalar quantities, so they only have magnitude.
>
> Displacement is a vector quantity, so it has direction as well as magnitude.

Speed

- The speed of an object is a measure of how fast it is travelling.
- It is a scalar quantity measured in metres per second (m/s).
- The speed that a person can walk, run or cycle depends on factors like age, fitness, terrain and distance.
- Some typical speeds can be seen below:

Method of Travel	Speed (m/s)	Method of Travel	Speed (m/s)
Walking	1.5	Motorway driving	30
Running	3	High speed trains	75
Cycling	6	Commercial aircraft	250
City driving	12	Speed of sound in air	330

- Most things (including sound) do not travel at a constant speed, so it is often the average speed over a period of time that is used.

distance travelled = speed × time

$$s = vt$$

Distance (s) is measured in metres (m).
Speed (v) is measured in metres per second (m/s).
Time (t) is measured in seconds (s).

Velocity

- Velocity is a vector quantity.
- It is the speed of an object in a given direction.

HT When travelling in a straight line, an object with constant speed also has constant velocity.

> **Key Point**
>
> Velocity is a vector quantity.

HT If the object is *not* travelling in a straight line, e.g. it is turning a corner:
- the speed can still be constant
- the velocity will change, because the direction has changed.

HT An object moving in a circle:
- is constantly changing direction, so it is constantly changing velocity
- is accelerating even if it is travelling at constant speed.

HT Orbiting planets are an example of this – it is the force of gravity that causes the acceleration.

Newton's First Law

- Newton's first law is often stated as: an object will remain in the same state of motion unless acted on by an external force.
- When the resultant force acting on an object is zero:
  - if the object is stationary, it remains stationary
  - if the object is moving, it continues to move at the same speed and in the same direction, i.e. at constant velocity.

  HT This tendency for objects to continue in the same state of motion is called inertia.

- The velocity (speed or direction) of an object will only change if there is a resultant force acting on it.
- For a car travelling at a steady speed, the driving force is balanced by the resistive forces.

Distance–Time Graphs

- A distance–time graph can be used to represent the motion of an object travelling in a straight line.
- The speed of the object is found from the gradient (slope) of the line.
- The graph on the right shows:
  1 A stationary object.
  2 An object moving at a constant speed of 2m/s.
  3 An object moving at a greater constant speed of 3m/s.

HT If an object is accelerating, the distance–time graph will be a curve.
HT For an accelerating object, its speed at a particular time is found by:
- drawing a tangent to the curve at the point in time
- working out the gradient of the tangent.

> ### Key Point
> **HT** An object travelling in a circle can have constant speed, but its velocity is still changing.

Constant Speed

Resistive force   Driving force

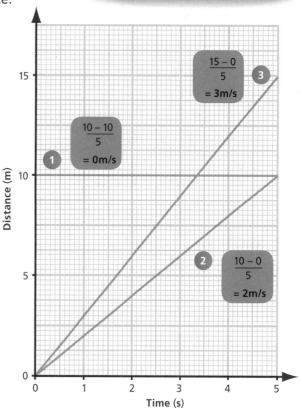

1    $\dfrac{10-10}{5}$ = 0m/s

2    $\dfrac{10-0}{5}$ = 2m/s

3    $\dfrac{15-0}{5}$ = 3m/s

Distance (m)

Time (s)

> ### Key Point
> A distance–time graph can be used to calculate speed.

> ### Key Words
> distance
> displacement
> speed
> velocity
> **HT** accelerating
> **HT** inertia

Quick Test

1. Describe the difference between distance and displacement.
2. Work out the speed of a car that travels 1km in 50 seconds.
3. What does the gradient of a distance–time graph show?

Forces and Acceleration

You must be able to:

- Apply Newton's second law to situations where objects are accelerating
- Estimate the magnitude of everyday accelerations
- Draw and interpret velocity–time graphs.

Acceleration

- The acceleration of an object is a measure of how quickly it speeds up, slows down or changes direction.

LEARN

$$\text{acceleration} = \frac{\text{change in velocity}}{\text{time taken}}$$

$$a = \frac{\Delta v}{t}$$

- When an object slows down, the change in velocity is negative, so it has a negative acceleration.
- Uniform acceleration can also be calculated using the equation:

$$(\text{final velocity})^2 - (\text{initial velocity})^2 = 2 \times \text{acceleration} \times \text{distance}$$

$$v^2 - u^2 = 2as$$

Velocity–Time Graphs

- The gradient of a velocity–time graph can be used to find the acceleration of an object.

HT The total distance travelled is equal to the area under the graph.

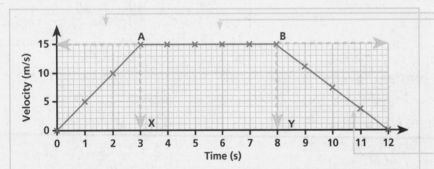

Distance travelled:

First section: $\frac{1}{2} \times 3 \times 15 = 22.5\text{m}$

Middle section: $15 \times 5 = 75\text{m}$

Final section: $\frac{1}{2} \times 4 \times 15 = 30\text{m}$

Total distance = $22.5 + 75 + 30$

$= 127.5\text{m}$

Key Point

Acceleration is a measure of rate of change of velocity.

Acceleration (a) is measured in metres per second squared (m/s²). Change in velocity (Δv) is found by subtracting initial velocity from final velocity ($v - u$) and is measured in metres per second (m/s). Time (t) is measured in seconds (s).

Final velocity (v) is measured in metres per second (m/s). Initial velocity (u) is measured in metres per second (m/s). Acceleration (a) is measured in metres per second squared (m/s²). Distance (s) is measured in metres (m).

The change in velocity is 15m/s over a 3 second period, so the acceleration is $\frac{15}{3}$ = 5m/s²

The velocity is constant, so the acceleration is zero.

The change in velocity is –15m/s over a 4 second period, so the acceleration is $\frac{-15}{4}$ = –3.75m/s². The negative value shows that the object is slowing down.

Break down the area under the graph into smaller shapes.

Add together all of the areas to find the total distance.

Newton's Second Law

- Newton's second law is often stated as: the acceleration of an object is proportional to the resultant force acting on the object and inversely proportional to the mass of the object, i.e.
    - if the resultant force is doubled, the acceleration will be doubled
    - if the mass is doubled, the acceleration will be halved.
- This law can be summarised with the equation:

$$\text{force} = \text{mass} \times \text{acceleration}$$

$$F = ma$$

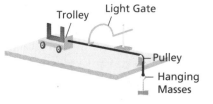

Force (F) is measured in newtons (N). Mass (m) is measured in kilograms (kg). Acceleration (a) is measured in metres per second squared (m/s^2).

HT Mass is a measure of inertia.

HT It describes how difficult it is to change the velocity of an object.

HT This inertial mass is given by the ratio of force over acceleration, i.e. $m = \dfrac{F}{a}$.

HT The larger the mass, the bigger the force needed to change the velocity.

REQUIRED PRACTICAL
Investigate the effect of varying the force and / or the mass on the acceleration of an object.

Sample Method	Considerations, Mistakes and Errors
1. Set up the equipment as shown. 2. Release the trolley and use light gates or a stopwatch to take the measurements needed to calculate acceleration. 3. Move 100g (1N) from the trolley onto the mass holder. 4. Repeat steps 2 and 3 until all the masses have been moved from the trolley onto the mass holder. If investigating the mass, keep the force constant by removing a mass from the trolley but not adding it to the holder.	• When changing the force it is important to keep the mass of the system constant. Masses are taken from the trolley to the holder. No extra masses are added. • Fast events often result in timing errors. Repeating results and finding a mean can help reduce the effect of these errors. • If the accelerating force is too low or the mass too high, then frictional effects will cause the results to be inaccurate.
Variables • The independent variable is the force or the mass. • The control variable is kept the same. In this case, the force if the mass is changed or the mass if the force is changed.	**Hazards and Risks** • The biggest hazard in this experiment is masses falling onto the experimenter's feet. To minimise this risk, masses should be kept to the minimum needed for a good range of results.

Trolley   Light Gate   Pulley   Hanging Masses

Key Point

Calculating a mean helps to reduce the effect of random errors.

A result that is accurate is close to the true value.

Quick Test

1. An object accelerates from 2m/s to 6m/s over a distance of 8m. Use the equation $v^2 - u^2 = 2as$ to find the acceleration of the object.
2. Comparing two velocity–time graphs, it can be seen that graph A is twice as steep as graph B. What does this indicate?
3. Why is it important to carry out repeat readings during an experiment?

Key Words

acceleration
gradient
proportional
inversely proportional
HT inertia

Terminal Velocity and Momentum

You must be able to:

- Draw and interpret diagrams showing how the forces on a falling object change as it approaches and reaches terminal velocity
- Give examples of Newton's third law in action
- **HT** Describe examples of conservation of momentum in collisions.

Terminal Velocity

- When an object falls through a fluid:
  - At first, the object accelerates due to the force of gravity.
  - As it speeds up, the resistive forces increase.
  - The resultant force reaches zero when the resistive forces balance the force of gravity. At this point the object will fall at a steady speed, called its terminal velocity.
- Near the Earth's surface, the acceleration due to gravity is 10m/s².
- The most common example of this is a skydiver:

  ❶ The skydiver accelerates due to the force of gravity.
  ❷ The skydiver experiences frictional force due to air resistance. Weight (W) is greater than the resistive forces (R), so the skydiver continues to accelerate.
  ❸ Speed and R increase and acceleration decreases.
  ❹ R increases until it is the same as W. At this point the resultant force is zero and the skydiver falls at terminal velocity.

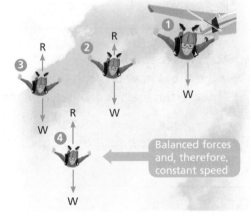

Balanced forces and, therefore, constant speed

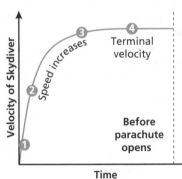

Newton's Third Law

- Newton's third law is often stated as: for every action there is an equal and opposite reaction.
- This means that whenever one object exerts a force on another, the other object exerts a force back.
- This reaction force is of the same type and is equal in size but opposite in direction.

A rocket pushes fuel backwards, which in turn pushes the rocket forwards.

HT Momentum

- All moving objects have momentum:

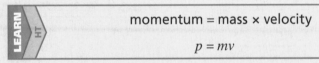

$$\text{momentum} = \text{mass} \times \text{velocity}$$

$$p = mv$$

Momentum (p) is measured in kilograms metres per second (kg m/s).
Mass (m) is measured in kilograms (kg).
Velocity (v) is measured in metres per second (m/s).

- When an unbalanced force acts on an object that is moving or able to move, a change in momentum occurs.
- Change in momentum can be calculated by substituting the equation for acceleration ($a = \frac{\Delta v}{t}$) into the equation for resultant force ($F = ma$):

$$F = \frac{m\Delta v}{\Delta t}$$

$m\Delta v$ is the change in momentum. Δt is the time over which the change takes place.

- This equation means: force equals rate of change of momentum.
- This is an important fact when considering many safety devices.
- These devices reduce the force by increasing the time over which the change of momentum takes place.
- For example, gymnasium crash mats cushion the impact of someone falling. They increase the time it takes for someone to come to rest when they fall onto the floor.

HT Conservation of Momentum

- In a closed system, the total momentum before an event is equal to the total momentum after the event.
- This conservation of momentum is most often referred to during collisions, but also applies to rockets and projectiles.

Two cars are travelling in the same direction along a road. Car A collides with the back of car B and they stick together.

Before

20m/s   9m/s

Car A mass 1200kg   Car B mass 1000kg

After

v m/s

Car A + Car B mass 2200kg

Calculate the velocity of the cars after the collision.

Momentum before collision = momentum A + momentum B
= (mass A × velocity of A) + (mass B × velocity of B)
= (1200 × 20) + (1000 × 9)
= 24000 + 9000 = 33 000kg m/s

Momentum after collision = 33 000kg m/s
Momentum after collision = (mass A + mass B) × (new combined velocity)

$$33\,000 = (1200 + 1000)v$$
$$33\,000 = 2200v$$
$$v = \frac{33\,000}{2200} = 15\text{m/s}$$

Quick Test

1. HT Calculate the momentum of a 2000kg car travelling at 20m/s.
2. Use Newton's third law and the idea of equal and opposite forces to explain how a fish propels itself through water.
3. HT a) Calculate the momentum of a horse and rider with a total mass of 600kg travelling at 8m/s.
   b) A motorcycle and rider travelling at 12m/s has the same momentum as the horse and rider in part a). Work out the combined mass of the motorcycle and rider.

HT Key Point

Momentum is the product of mass and velocity.

Force is the rate of change of momentum.

Conservation of Momentum

Recoil

Explosion

Rocket propulsion

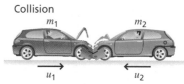

Collision

m_1 m_2

u_1 u_2

Start by calculating the momentum before the collision.

Remember, momentum is conserved so: momentum before collision = momentum after collision.

Substitute in the values for momentum and mass.

Rearrange the equation to find the velocity.

Key Point

In a closed system, momentum is conserved.

Key Words

terminal velocity
HT momentum
HT collision

Stopping and Braking

You must be able to:

- Interpret graphs that relate speed to stopping distance
- Describe factors affecting reaction time and braking distance
- Explain the dangers of large decelerations.

Stopping Distance

- The stopping distance of a vehicle depends on:
  - the thinking distance (the distance travelled during the driver's reaction time)
  - the braking distance (the distance travelled under the braking force).

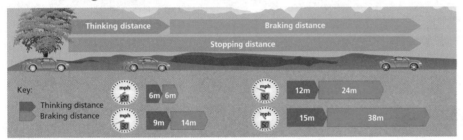

- For a given braking force: the greater the speed of the vehicle, the longer the stopping distance.

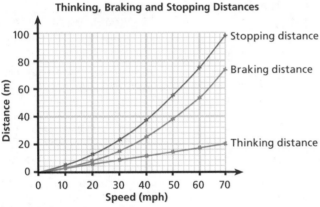

Thinking, Braking and Stopping Distances

- From the graph it can be seen that:
  - the thinking distance is directly proportional to speed
  - doubling the speed increases the braking distance by a factor of 4.

Key Point

Stopping distance is the sum of the thinking distance and the braking distance.

Reaction Time

- Reaction times vary from person to person, but the typical human reaction time is in the range of 0.2–0.9 seconds.
- This means that a car travelling at 30m/s (≈70mph) will travel between 12 and 27 metres before the person even begins to brake.
- This reaction time can be affected by tiredness, drugs and alcohol.
- Distractions, e.g. mobile phone use, also affect a person's ability to react.
- To measure reaction time, use lights or sounds as a 'start' signal and an electronic timer to measure how long someone takes to react.

Key Point

Reaction time can be affected by alcohol, drugs, fatigue and distractions.

- In the classroom, reaction times can be measured by dropping a ruler vertically and catching it as it falls.
- The distance the ruler falls through a person's fingers can be used to calculate the time it took them to react.

Factors Affecting Braking Distance

- The braking distance of a vehicle can be affected by the condition of the road, the vehicle and the weather.
- Adverse weather conditions include wet or icy / snowy roads.
- Vehicle condition includes factors such as worn brakes or tyres and over-inflated or under-inflated tyres.
- To stop a vehicle, the brakes need to apply a force to the wheels.
- The greater the braking force, the greater the deceleration of the vehicle.
- Work done by this frictional force transfers the kinetic energy of the vehicle into heat energy, increasing the temperature of the brakes.
- If the braking force is too large, the brakes may overheat or the tyres may lose traction on the road resulting in the car skidding.
- Overheating and loss of traction are more likely to occur if the brakes or tyres are in poor condition.
- When a vehicle is travelling faster, it needs a larger braking force to be able to stop it in a certain distance.

HT To find the size of the braking force required, the equation for work done can be used:

> work done (kinetic energy) = force × distance (braking distance)

HT The table below shows the size of the braking force involved in two different braking situations:

Vehicle	Mass	Speed	Kinetic Energy	Braking Distance	Force
Car	1500kg	City ≈ 15m/s	168 750J	50m	3 375N
Lorry	7500kg	Motorway ≈ 30m/s	3 375 000J	50m	67 500N

The symbol ≈ means approximately equal to. Sometimes a single ~ is used to mean the same thing.

- For a given braking distance:
  - doubling the mass doubles the force required
  - doubling the speed quadruples the force required.

> **Quick Test**
>
> 1. Explain what effect talking on a mobile phone might have on the stopping distance of a car.
> 2. **HT** Estimate the braking force needed to stop a 750kg motor cycle travelling at city speeds over a distance of 25m.
> 3. Describe a simple experiment to measure the reaction time of a person.

> **Key Words**
>
> braking distance
> deceleration

Forces – An Introduction

1 The mass of an object is a scalar quantity, but the weight of an object is a vector quantity.

Explain what is meant by this statement and the link between mass and weight. [4]

2 **Figure 1** represents the forces acting on an object.

Figure 1

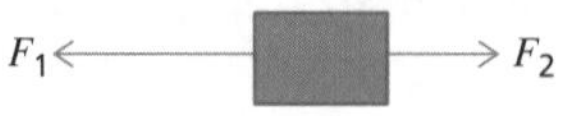

a) Compare the forces F_1 and F_2. [2]

$F_1 \leftarrow$ ▭ $\rightarrow F_2$

b) The force F_2 increases until it is of equal magnitude to F_1.

What will be the magnitude of the resultant force? [1]

3 HT **Figure 2** represents the resultant force acting on an object.

Figure 2

Draw the horizontal and vertical components of the arrow onto **Figure 2**.

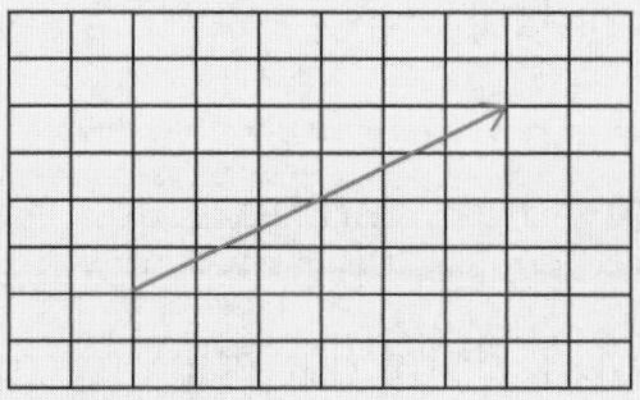

[1]

Total Marks _____ / 8

Forces in Action

1 A car travelling along a straight, level road at 30mph has 150kJ of kinetic energy.
The driver applies the brakes and comes to a complete stop in 50m.

a) What happens to the temperature of the brakes during braking? [1]

b) How much work is done by the brakes to stop the car? [1]

c) Calculate the braking force applied by the brakes. [2]

2 A wheelbarrow is loaded with 60kg of soil.
The load is a distance of 20cm from the wheel, which acts as the pivot point when the barrow is lifted.

a) Show that the soil weighs around 590N (g = 10N/kg). [2]

b) The handles of the wheel barrow are positioned 60cm from the wheel.

Estimate the force needed to use the wheelbarrow to lift the soil. [2]

Total Marks _____ / 8

Pressure and Pressure Differences

1. Describe the cause of pressure in fluids. [2]

2. Write down the formula that links pressure, force and area. [1]

3. In a hydraulic system, the cross-sectional area of the left-hand piston is $\frac{1}{10}$ of the cross-sectional area of the right-hand piston.

   If the left-hand piston is pushed down with a force of 100N, what force will be transferred to the right-hand piston? [2]

4. HT For this question you will need the following information:

   > pressure =
   >   height of the column × density of the liquid × gravitational field strength
   > Gravitational field strength is 10N/kg.
   > Density of water is 1000kg/m³.

   a) A submarine descends from the surface to a depth of 100m.

      Work out the increase in pressure. [3]

   b) A swimming pool is 3m deep.

      Calculate the pressure change experienced by swimmers when they swim from the bottom to the surface. [3]

5. A small boat floating on a lake experiences an upthrust 2597N.
   (Use 10N/kg for the gravitational field strength.)

   a) Calculate the mass of the boat to the nearest kilogram. [2]

   The boat is loaded with an additional 100kg. It sits lower in the water but still floats.

   b) What is the new total mass of the boat? [1]

   c) Calculate the new value for the upthrust. [2]

   d) Work out the volume of water displaced by the boat.
      Density of water is 1000kg/m³.

      $$\text{density} = \frac{\text{mass}}{\text{volume}}$$ [4]

   Total Marks _____ / 20

Practice Questions

Forces and Motion

1 A hiker travels north for 2 miles, east for 1 mile and then south again for 2 miles.

    a) What is the total distance travelled? [1]

    b) What is the final displacement? [2]

2 The length of the race track at Silverstone is 3.65 miles.
A Formula One race is 52 laps of the race track.
The winner of the 2015 race completed the race in approximately 1 hour and 30 minutes.

    a) Calculate the following:

        i) The total distance travelled by one car during the race. [1]

        ii) The displacement of the car at the end of the race, after completing 52 laps. [1]

        iii) The average speed (in mph) of the winning car over the entire race. [2]

    b) At some points in the race, the cars will be travelling at constant speed but their velocity will be changing.

    Explain how this can be true and where on the track this might occur. [3]

3 **Figure 1** is a distance–time graph.

    a) How long is the object stationary for in total? [1]

    b) During which part of the journey is the object travelling at the greatest speed. [1]

    c) What is the speed between points **A** and **B**? [2]

    d) If the graph contained curved lines, what would the curved sections indicate? [1]

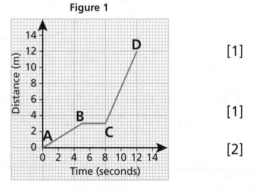

Figure 1

Total Marks _____ / 15

Forces and Acceleration

1 A car travelling at 10m/s accelerates at a constant rate of 4m/s² over a distance of 100m.

Use the formula $v^2 - u^2 = 2as$ to work out the final velocity reached by the car. [4]

2 Table 1 shows the velocity of a remote-controlled car every two seconds for a 10 second period.

Time (s)	Velocity (m/s)
0	0
2	4
4	8
6	8
8	8
10	0

a) Plot a velocity–time graph for the 10 second period. [3]

b) Using your graph, describe the motion of the car between 5 seconds and 7 seconds. [1]

c) Calculate the acceleration of the car during the first 4 seconds. [2]

d) HT From your graph, calculate the total distance travelled. [3]

Total Marks _____ / 13

Terminal Velocity and Momentum

1 A falling object takes 0.4 seconds to accelerate from rest to a speed of 4m/s. Assuming that no other forces act on the object, show that the acceleration due to gravity is 10m/s^2. [3]

2 A helicopter weighs 25 000N.

a) What is the size of the upwards force that must act on the helicopter for it to remain hovering in a stationary position? [1]

b) Use Newton's third law to explain how the helicopter produces the force required in part a). [3]

Total Marks _____ / 7

Stopping and Braking

1 The airbags in a car inflate instantly in a crash and then deflate when a person hits them.

Explain how this helps to prevent injuries in terms of force, time and change in momentum. [4]

2 Give **three** factors that would have a negative effect on the reaction time of a driver. [3]

3 HT A lorry has a mass of 5000kg and is travelling at 12m/s.
The driver applies the brakes and takes 5 seconds to come to a complete stop.

Calculate the braking force applied.

$$\text{force} = \frac{\text{change in momentum}}{\text{time taken}}$$

[2]

Total Marks _____ / 9

Energy Stores and Transfers

You must be able to:

- Describe changes in the way energy is stored when a system changes
- Calculate the energy changes involved when a system changes
- Describe what is meant by internal energy
- Use specific heat capacity in calculations
- Describe how to investigate the specific heat capacity of materials.

Energy Stores and Systems

- A **system** is an object or group of objects.
- When a system changes, there are changes in the way **energy** is stored, e.g. a roller coaster transfers energy between gravitational and kinetic energy; an electric kettle transfers electrical energy to thermal energy.
- Diagrams are a useful way to illustrate how the energy is redistributed when a system is changed.
- The Sankey diagram below shows that the light bulb redistributes the electrical energy as heat and light.

On most rollercoasters, the cars start **high up** with a lot of **gravitational potential energy.**

As the cars **drop**, the **gravitational potential energy** is gradually **transferred** into **kinetic energy.**

Calculating Energy Changes

- You need to be able to calculate the amount of energy associated with a moving object, an object raised above ground level and a stretched spring.
- The **kinetic energy** of a moving object can be calculated by:

LEARN

$$\text{kinetic energy} = 0.5 \times \text{mass} \times (\text{speed})^2$$

$$E_k = \tfrac{1}{2}mv^2$$

Sankey Diagram of an Incandescent Light Bulb

Light energy
10J

Electrical energy
100J

Heat energy
90J

- The **gravitational potential energy (GPE)** gained by raising an object above ground level can be calculated with the equation:

LEARN

$$\text{gravitational potential energy} = \text{mass} \times \text{gravitational field strength} \times \text{height}$$

$$E_p = mgh$$

Kinetic energy (E_k) is measured in joules (J).
Mass (m) is measured in kilograms (kg).
Speed (v) is measured in metres per second (m/s).

- The amount of **elastic energy** stored in a stretched or compressed spring can be found with the equation:

$$\text{elastic potential energy} = 0.5 \times \text{spring constant} \times (\text{extension})^2$$

$$E_e = \tfrac{1}{2}ke^2$$

Gravitational potential energy (E_p) is measured in joules (J).
Mass (m) is measured in kilograms (kg).
Gravitational field strength (g) is measured in newtons per kilogram (N/kg).
Height (h) is measured in metres (m).

Specific Heat Capacity and Internal Energy

- **Internal energy** is the total kinetic and potential energy of all the particles that make up a system.
- Doing work on a system increases the energy stored in a system.

Elastic potential energy (E_e) is measured in joules (J).
Spring constant (k) is measured in newtons per metre (N/m).
Extension (e) in metres (m).

- Heating changes the energy stored in a system by increasing the energy of the particles within it.
- As the energy increases, this will either increase the temperature or produce a change of state (see pages 84–85).
- If the temperature increases, the increase depends on:
  - the mass of the substance heated
  - what the substance is
  - the energy input.
- The **specific heat capacity** of a substance is the amount of energy required to raise the temperature of one kilogram of the substance by one degree Celsius.

(see pages 84–85)

> change in thermal energy =
> mass × specific heat capacity × temperature change
>
> $$\Delta E = mc\Delta\theta$$

Calculate the increase in temperature when 2100J of energy is provided to 100g of water. The specific heat capacity of water is 4200 J/kg °C.

$$2100 = 0.1 \times 4200 \times \Delta\theta$$
$$\Delta\theta = \frac{2100}{(4200 \times 0.1)} = 5°C$$

Key Point

The amount of energy stored can also be found by the amount of work done, so you need to look at the information you are given. If you have not been given the information needed to use the elastic potential energy equation, but have been given force and distance, then use work done = force × distance

Change in thermal energy (ΔE) is measured in joules (J).
Mass (m) is measured in kilograms (kg).
Specific heat capacity (c) is measured in joules per kilogram per degree Celsius (J/kg °C).
Temperature change ($\Delta\theta$) is measured in degrees Celsius (°C).

Substitute in the given values into the equation $\Delta E = mc\Delta\theta$. Make sure they are in the correct units, i.e. 100g = 0.1kg.

Rearrange the equation to find the temperature change ($\Delta\theta$).

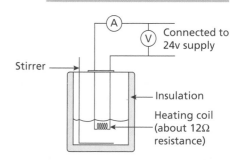
Connected to 24v supply
Stirrer
Insulation
Heating coil (about 12Ω resistance)

REQUIRED PRACTICAL

Investigate the specific heat capacity of materials, linking the decrease of one energy store (or work done) to the increase in temperature and subsequent increase in thermal energy stored.

Sample Method
1. Set up the apparatus as shown.
2. Measure the start temperature.
3. Switch on the electric heater for 1 min.
4. Measure the end temperature.
5. Measure the voltage and current to find the power.
6. Repeat for different liquids.
7. Calculate the specific heat capacity.
8. Compare your results to another group's. If they get similar answers the experiment is **reproducible**.

Considerations, Mistakes and Errors
- The energy provided by the heater is calculated as power × time. However, it could also be found using a joulemeter.
- The specific heat capacity is calculated from the energy provided, the mass of the liquid and the temperature change.
- If the temperature rise is too high, energy loss to the surroundings will affect the results.

Variables
- The independent variable is the type of liquid.
- The dependent variable is the temperature.
- Control variables are the amount of liquid used and energy provided.

Hazards and Risks
- The electric heater could be very hot so you must *not* touch it directly.
- If the liquids become hot they could boil and spit, so safety goggles must be worn and the heater should not be left on for longer than is necessary.

Quick Test

1. Use a Sankey diagram to illustrate the energy changes involved when a diesel powered crane is used to lift an object.
2. Work out the energy required to heat 1.5kg of water from 20°C to 100°C. (c = 4200J/kg °C)
3. Explain what is meant by 'reproducible' and why it is important.

Key Words

system
energy
kinetic energy
gravitational potential energy (GPE)
elastic energy
internal energy
specific heat capacity
reproducible

Energy Transfers and Resources

You must be able to:

- Describe examples of energy transfers in a closed system
- Describe how thermal conductivity affects rate of cooling
- Distinguish between renewable and non-renewable energy resources.

Energy Transfers

- When looking at energy transfer, there are two key points:
  - Energy can be transferred usefully, stored or dissipated (spread out to the surroundings).
  - Energy cannot be created or destroyed.
- In a closed system the total energy never changes, but it can be transferred from one store to another.
- For example, when an electricity-powered lift raises the lift carriage:
  - it transfers electrical energy into gravitational potential energy
  - some energy is dissipated into the surroundings as heat and sound
  - this wasted energy is no longer available for useful transfers.
- Wasted energy is caused by unwanted energy transfers.
- These unwanted transfers can be reduced in several ways:
  - lubrication – reduces the friction that produces heat
  - tightening any loose parts – prevents unwanted vibration that wastes energy as sound
  - thermal insulation – reduces heat loss.
- A system isn't always a single device; it could be an entire building.
- A building wastes energy when it loses heat to the surroundings, causing it to cool down.
- The rate of cooling depends on the thickness and thermal conductivity of the walls.
- Thin walls with high thermal conductivity will conduct heat the quickest and the building will cool down rapidly.

> **Key Point**
>
> An anomalous result is one that doesn't fit the pattern. These:
> - should be looked at to try and determine the cause
> - should be ignored when plotting graphs
> - should not be included when calculating averages.

REQUIRED PRACTICAL
Investigate the effectiveness of different materials as thermal insulators.

Sample Method	**Considerations, Mistakes and Errors**
A simple method is to compare different methods of insulation: 1. Take four test tubes and wrap each one in a different type of insulation. 2. Fill each test tube with hot water and measure the start temperature of each one. 3. Start the stopwatch and record the temperature every minute for 10 mins. 4. Plot the results on a graph of time against temperature.	• Time and temperature are examples of continuous data, so a line graph should be used. This will allow any patterns and anomalous results to be easily spotted. • A cooling curve should be a smooth curved line. If the temperature goes up or down suddenly, this will not fit the pattern of the curve and should be ignored as an anomalous result.
Variables • The independent variable is the type of insulation. • The dependent variable is temperature. • The control variables are the times at which the temperature is measured, the volume of water in each test tube and the thickness of insulating material.	**Hazards and Risks** • The main hazard is the hot water, which scalds, so care must be taken when pouring the water into the test tubes. • If water is spilt on the insulating material it will affect the results. • The test tubes may still be hot even after 10 minutes, so these should be allowed to cool before disposal.

National and Global Energy Resources

- The main uses for energy resources are transport, electricity generation and heating.
- All energy resources fall into one or two categories:
  - **renewable** energy resources, which can be replenished
  - non-renewable energy resources, which will eventually run out.

Category	Energy Resource	Main Uses	Environmental Impacts, Ethics, Reliability and Other Information
Renewable	Biofuel	Transport and electricity generation	Large areas of land are needed for growing fuel crops. This can be at the expense of food crops in poorer countries.
Renewable	Wind	Electricity generation	Does not provide a constant source of energy. Turbines can be noisy / dangerous to birds. Some people think they ruin the appearance of the countryside.
Renewable	Water (hydro-electricity)	Electricity generation	Requires large areas of land to be flooded, altering ecosystems and displacing the people that live there.
Renewable	Geothermal	Electricity generation and heating	Only available in a limited number of places where hot rocks can be found close to the surface, e.g. Iceland.
Renewable	Tidal	Electricity generation	Variations in tides affect output. Have a high initial set-up cost. Can alter habitats / cause problems for shipping.
Renewable	Solar	Electricity generation and some heating	Depends on light intensity, so no power produced at night. High cost in relation to power output
Renewable	Water waves	Electricity generation	Output depends on waves, so can be unreliable. Can alter habitats.
Non-renewable	Nuclear fuel	Electricity generation and some military transport	Produces radioactive waste but no other emissions. Costly to build and decommission. Reliable output.
Non-renewable	Coal	Electricity generation, heating and some transport	Burning produces greenhouse gases (CO_2) and contributes to acid rain (SO_2). Reliable output.
Non-renewable	Oil	Transport and heating	Reliable output. Provides a compact source of energy for transport. Burning produces CO_2, NO_2 and SO_2. Serious environmental damage if spilt.
Non-renewable	Gas	Electricity generation, heating and some transport	Reliable output. Burning produces CO_2 but *not* SO_2.

Quick Test

1. A 20W light bulb transfers 15W usefully as light energy. What happens to the energy that is not transferred as light?
2. Describe one ethical consideration of hydroelectric power.

Key Words

transferred
dissipated
conductivity
anomalous
renewable

Waves and Wave Properties

You must be able to:

- Describe the difference between transverse and longitudinal waves
- Describe evidence that waves transfer energy not matter
- Describe waves in terms of amplitude, wavelength, frequency and period
- Explain how wave speed, frequency and wavelength are linked.

Transverse and Longitudinal Waves

- There are two types of wave: transverse and longitudinal.
- All waves transfer energy from one place to another.
- For example, if a stone is dropped into a pond, ripples travel outwards carrying the energy. The water does not travel outwards (otherwise it would leave a hole in the middle).
- The particles that make up a wave oscillate (vibrate) about a fixed point. In doing so, they pass the energy on to the next particles, which also oscillate, and so on.
- The energy moves along, but the matter remains.
- In a transverse wave, e.g. water wave, the oscillations are perpendicular (at right-angles) to the direction of energy transfer.
- This can be demonstrated by moving a rope or slinky spring up and down vertically – the wave then moves horizontally.
- In a longitudinal wave, e.g. sound wave, the oscillations are parallel to the direction of energy transfer.
- This can be demonstrated by moving a slinky spring back and forward horizontally – the wave also moves horizontally.

Properties of Waves

- All waves have a:
  - **frequency** – the number of waves passing a fixed point per second, measured in hertz (Hz)
  - **amplitude** – the maximum displacement that any particle achieves from its undisturbed position in metres (m)
  - **wavelength** – the distance from one point on a wave to the equivalent point on the next wave in metres (m)
  - **period** – the time taken for one complete oscillation in seconds (s).
- When observing waves (e.g. water waves):
  - amplitude is seen as the wave height
  - the period is the time taken for one complete wave to pass a fixed point.

Key Point

Waves transfer energy **not** matter.

Transverse Waves

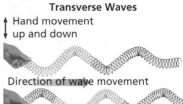

Hand movement
up and down

Direction of wave movement

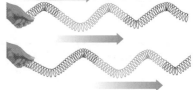

Longitudinal Waves

Hand movement
in and out

Compression   Rarefaction

Direction of wave movement

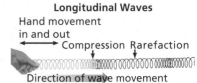

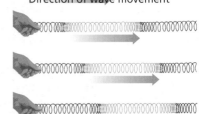

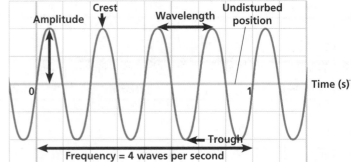

$$period = \frac{1}{frequency}$$

$$T = \frac{1}{f}$$

Period (T) is measured in seconds (s).
Frequency (f) is measured in hertz (Hz).

- Amplitude indicates the amount of energy a wave is carrying – the more energy, the higher the amplitude.

Wave Speed

- The speed of a wave is the speed at which the energy is transferred (or the wave moves).
- It is a measure of how far the wave moves in one second and can be found with 'the wave equation':

wave speed = frequency × wavelength

$$v = f\lambda$$

Wave speed (v) is measured in metres per second (m/s).
Frequency (f) is measured in hertz (Hz).
Wavelength (λ) is measured in metres (m).

- Ripples on the surface of water are slow enough that their speed can be measured by direct observation and timing with a stopwatch.

REQUIRED PRACTICAL
Identify the suitability of apparatus to measure the frequency, wavelength and speed of waves in a ripple tank.

Sample Method	**Considerations, Mistakes and Errors**
1. Time how long it takes one wave to travel the length of the tank. Use this to calculate wave speed using $speed = \dfrac{distance}{time}$. 2. To find the frequency, count the number of waves passing a fixed point in a second. 3. Estimate the wavelength by using a ruler to measure the peak-to-peak distance as the waves travel. 4. Use a stroboscope to make the same measurements and compare the results.	• Using a stroboscope can significantly improve the accuracy of measurements. • By projecting a shadow of the waves onto a screen below the stroboscope, flash speed can be adjusted to make the waves appear stationary. This makes wavelength measurements much more accurate. • For high frequencies that are difficult to count, this can be used with the wave speed measurement to calculate the frequency using $f = \dfrac{v}{\lambda}$.
Variables	**Hazards and Risks**
• The key control variable is water depth. It is important to ensure that the depth of the water is kept constant across the tank as, for a given frequency, the depth will affect the speed and wavelength.	• When using a stroboscope there is a risk to people with photo-sensitive epilepsy. It is important to check that there are no at-risk people involved in the experiment or in the area.

- As waves are transmitted from one medium to another, their speed and, therefore, their wavelength changes, e.g. water waves travelling from deep to shallow water or sound travelling from air into water.
- The frequency does not change because the same number of waves is still being produced by the source per second.
- Because all waves obey the wave equation, the speed and wavelength are directly proportional:
  - doubling the speed, doubles the wavelength
  - halving the speed, halves the wavelength.

Key Words

transverse
longitudinal
oscillate
frequency
amplitude
wavelength
period

Quick Test

1. In what direction are the oscillations in a transverse wave?
2. Work out the period of a wave with a frequency of 20Hz.
3. Describe a method for measuring wave speed.

Reflection, Refraction and Sound

You must be able to:

- Construct ray diagrams to illustrate reflection of a wave at a surface
- Construct ray diagrams to illustrate refraction of a wave at a boundary
- **HT** Use wave front diagrams to explain refraction
- Describe an experiment to investigate reflection or refraction
- **HT** Explain how sound waves are produced, transmitted and detected.

Reflection

- When waves reach a boundary between one medium (material) and another, they can be reflected, refracted, absorbed or transmitted.
- Ray diagrams can be used to show what happens.
- When constructing a ray diagram:
  - Rays must be drawn with a ruler.
  - Each straight section of ray should have a single arrow drawn on it to indicate the direction of movement.
  - Where the ray meets the boundary, a 'normal' should be drawn at right-angles to the boundary.
  - All relevant angles should be labelled.
- When waves are reflected at a surface, the angle of incidence is equal to the angle of reflection.

Refraction

- When a wave passes from one medium into another, it can be refracted and change direction. The direction of refraction depends on:
  - the angle at which the wave hits the boundary
  - the materials involved.
- For light rays, the way in which a material affects refraction is called its refractive index.
- When light travels:
  - from a material with a low refractive index to one with a higher refractive index, it bends towards the normal
  - from a material with a high refractive index to one with a lower refractive index, it bends away from the normal.

HT Refraction is due to the difference in the wave speed in the different media.

HT When a light wave enters, at an angle, a medium in which it travels slower:
  - the first part of the light wave to enter the medium slows down
  - the rest of the wave continues at the higher speed
  - this causes the wave to change direction, towards the normal.

HT You need to be able to show this using a wave front diagram like the one alongside, which is for water waves.

Light Reflected by a Plane Mirror

Object — Incident ray (travelling towards mirror)

Eye — Reflected ray (travelling away from mirror)

Normal

i   r

Plane mirror

Point of incidence

→ = Light ray   i = Angle of incidence   r = Angle of reflection

Key Point

A common error is to draw the ray refracting onto or even past the normal. This *never* happens and will lose marks in the exam.

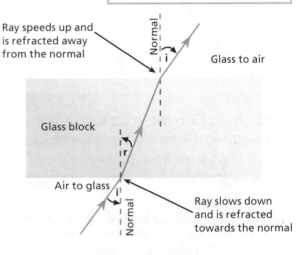

Ray speeds up and is refracted away from the normal

Normal

i

Glass to air

Glass block

r

Air to glass

i

Normal

Ray slows down and is refracted towards the normal

Refraction

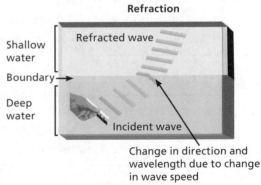

Shallow water

Refracted wave

Boundary

Deep water

Incident wave

Change in direction and wavelength due to change in wave speed

REQUIRED PRACTICAL	
Investigate reflection and refraction.	
Sample Method	**Considerations, Mistakes and Errors**
1. Set up the equipment as shown. 2. Draw around the semi-circular block. 3. Mark the position of the light ray at the start, at the end, and where it enters and exits the block. 4. Remove the block and connect the marks to show the light rays. 5. Add normal lines and measure the angles of incidence and refraction. 6. Repeat on a new piece of paper for a range of incidence angles. 7. Repeat the experiment with blocks made of different transparent materials.	• It is important to keep the light ray as narrow as possible and the incident at the exact centre of the flat side of the block. If the ray is too wide or off centre, it can lead to inaccurate measurements. • It is also essential that the block is removed before trying to measure the angles. If the protractor does not line up correctly, it will create a zero error, i.e. the protractor would read a value other than zero, when the angle was zero, making all other readings incorrect.
Variables	**Hazards and Risks**
• The independent variable is the angle of incidence. • The dependent variable is the angle of refraction. • The control variable is the material the block is made from.	• If the light box is left on for a long time, the housing close to the bulb can become hot enough to burn. The ray box should be turned off when not in use and care should be taken when moving it.

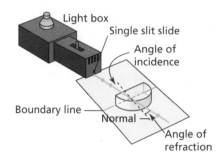

Light box
Single slit slide
Angle of incidence
Boundary line
Normal
Angle of refraction

HT Sound Waves

- Sound waves, like any wave, have frequency, amplitude and wavelength.
- The amplitude of a sound wave relates to the loudness.
- The frequency and wavelength of a sound wave relate to the pitch – the higher the frequency, the higher the pitch.
- The normal range of human hearing is from 20Hz to 20kHz (20 000Hz).
- Sound can travel through liquids and solids, as well as air.
- Sound, in any medium, is due to vibration of the particles that make up the medium.
- In a solid, these oscillations can cause the entire object to vibrate with the same frequency as the sound wave.
- The conversion of sound waves to vibrations only occurs over a limited range of frequencies. The range of frequencies converted is dependent on the structure of the object.
- Within the ear, sound waves cause the ear drum and other structures to vibrate and it is this vibration that is heard as sound.
- The limited range of conversion is what limits human hearing.
- You need to be able to give examples of sound waves being converted to vibration, e.g. in the ear drum, by a microphone or a glass being shattered by an opera singer.

> **Quick Test**
>
> 1. On a ray diagram what is the 'normal'?
> 2. A light wave travels from one medium to another. As it crosses the boundary, it speeds up. In which direction does the light ray turn?
> 3. HT What is the highest frequency sound that humans can normally hear?

Waves for Detection and Exploration

You must be able to:

HT Describe how ultrasound can be used in medicine and industry

HT Explain how echo sounding is used to detect objects in deep water and to measure water depth

HT Describe the different types of seismic wave

HT Explain how the study of seismic waves provided evidence about the structure of the planet Earth.

Ultrasound

- **Ultrasonic** waves have a frequency greater than 20kHz, so they cannot be heard by humans.
- When an ultrasonic wave meets a boundary between two different media it is partially reflected.
- It is possible to determine how far away a boundary is by measuring the time taken for reflected ultrasonic waves to return to a detector.
- Uses in industry include detecting defects in materials without cutting into them. These defects could be manufacturing faults (e.g. cracks and air bubbles) or damage (e.g. corrosion).
- Uses in medicine include pre-natal scanning, detection of kidney stones and tumours, and producing images of damaged ligaments and muscles.

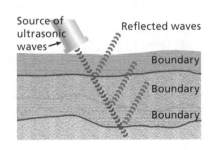

Echo Sounding

- **Echo sounding**, or sonar, is the use of ultrasonic waves for detecting objects in deep water and measuring the depth of water.
- It involves sending an ultrasound pulse into the water, which is then reflected back when it hits a surface.
- The time between the pulse being sent and the reflection being detected is used to calculate the distance travelled by the sound wave:

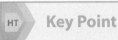

> distance = speed × time

- The speed is the speed of sound in water (1500m/s).
- This will find the total distance travelled by the pulse, which is then divided by two to find the depth of the water.

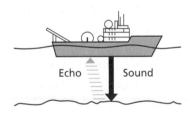

Echo ⎯ Sound

A ship sends an ultrasound pulse into the sea.
The reflection from the seabed is detected 0.5 seconds later.
Work out the depth of the water.

distance = speed × time
$$= 1500 \times 0.5 = 750m$$
water depth $= \dfrac{750}{2} = 375m$

> HT **Key Point**
>
> Ultrasound is sound with a frequency too high for humans to hear.

> HT **Key Point**
>
> You need to be able to explain everyday and technological uses of science like this.

> This is the total distance travelled.

> Don't forget to divide by 2 to find the depth of water.

Seismic Waves

- Two types of **seismic** wave are produced during an earthquake:
    - **P-waves** (primary)
    - **S-waves** (secondary).
- P-waves:
    - are longitudinal waves
    - travel at the speed of sound and are twice as fast as S-waves
    - travel at different speeds through solids and liquids.
- S-waves:
    - are transverse waves
    - are not able to travel through liquids.
- When seismic waves are produced, the difference in time between the arrival of P-waves and S-waves at different detectors can provide evidence about the location of the earthquake and the material they have travelled through.

The Structure of the Earth

- During an earthquake:
    - seismic waves travel outwards from the earthquake and are capable of travelling all the way through the Earth
    - seismic waves travel in a curved path through the Earth, due to the Earth increasing in density with depth
    - detectors placed around the Earth measure when and where the different waves arrive.
- There are two key pieces of evidence, which come from the S-wave shadow zone and the P-wave shadow zone.
- S-wave shadow zone:
    - S-waves are not able to travel through the liquid outer core of the Earth.
    - This results in a large shadow zone on the opposite side of the Earth to where the earthquake originated.
    - This shadow zone provides evidence of the size of the Earth's core.
- P-wave shadow zone:
    - P- waves are able to travel through the liquid outer core.
    - However, they are refracted at the boundary between the semi-solid mantle and the liquid outer core.
    - They are refracted again at the boundary between the liquid outer core and solid inner core.
    - These refractions result in P-wave shadow zones.
    - The study of these shadow zones is used to determine the size and composition of the inner and outer core.

> **HT** **Key Point**

Ultrasonic imaging and echo sounding both work by detecting the time between a transmitted pulse and a reflected pulse.

The Pattern of P-Wave Paths Through Earth's Interior

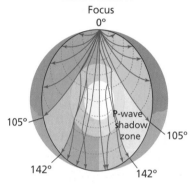

The Pattern of S-Wave Paths Through Earth's Interior

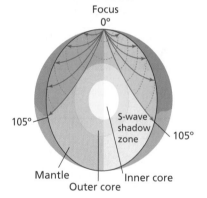

> **HT** **Key Point**

There are two types of seismic waves: P-waves and S-waves.

Seismic waves have provided evidence about the structure of the Earth.

> **HT** **Quick Test**

1. A ship sends a sonar pulse. The pulse reflected from the seabed is detected 1.2 seconds later. How deep is the sea? (Speed of sound in water = 1500m/s.)
2. Describe **two** key differences between P-waves and S-waves.
3. Explain the origin of the S-wave shadow.

> **HT** **Key Words**

ultrasonic
echo sounding
seismic
P-waves
S-waves

The Electromagnetic Spectrum

You must be able to:

- Recall the order of waves in the electromagnetic spectrum
- Give examples of electromagnetic waves transferring energy
- **HT** Explain why different electromagnetic waves are suitable for particular applications
- Describe how electromagnetic waves can be produced
- Explain the risks and consequences of radiation exposure.

Electromagnetic Waves

- **Electromagnetic (EM) waves** are transverse waves.
- All types of electromagnetic wave travel at the same velocity (the speed of light) in air or a vacuum.
- The electromagnetic spectrum extends from low frequency, low energy waves to high frequency, high energy waves.
- Human eyes are only capable of detecting visible light, i.e. a very limited range of electromagnetic waves.
- **HT** The wavelength of an electromagnetic wave affects how it is absorbed, transmitted, reflected or refracted by different substances. This affects its uses.

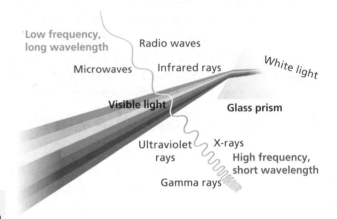

Low frequency, long wavelength — Radio waves — Microwaves — Infrared rays — Visible light — White light — Glass prism — Ultraviolet rays — X-rays — Gamma rays — High frequency, short wavelength

Uses and Applications of EM Waves

EM Wave	Uses	**HT** Explanations
Radio waves	Television, radio, Bluetooth	• Radio waves are low energy waves and, therefore, not harmful, making them ideal for radio transmission.
Microwaves	Satellite communications, cooking food	• Microwaves travel in straight lines through the atmosphere. • This makes them ideal for transmitting signals to satellites in orbit and transmitting them back down to receivers.
Infrared waves	Electrical heaters, cooking food, infrared cameras	• Electrical heaters, grills, toasters, etc. glow red hot as the electricity flows through them. • This transmits infrared energy that is absorbed by the food and converted back into heat.
Visible light	Fibre optic communications	• Visible light travels down optical fibres from one end to the other without being lost through the sides.
Ultraviolet waves	Energy efficient light bulbs, security marking, sunbeds	• In energy efficient light bulbs, UV waves are produced by the gas in the bulb when it is excited by the electric current. • These UV waves are absorbed by the coating on the bulb, which fluoresces giving off visible light.
X-rays	Medical imaging and treatments	• X-rays are able to penetrate soft tissue but not bone. • A photographic plate behind a person will show shadows where bones are.
Gamma rays	Sterilising food, treatment of tumours	• Gamma rays are the most energetic of all electromagnetic waves and can be used to destroy bacteria and tumours.

REQUIRED PRACTICAL	
Investigate how the amount of infrared radiation absorbed or radiated by a surface depends on the nature of that surface.	

Sample Method	**Considerations, Mistakes and Errors**
1. Take four boiling tubes each painted a different colour: matt black, gloss black, white and silvered. 2. Pour hot water into each boiling tube. 3. Measure and record the start temperature of each tube. 4. Measure the temperature of each tube every minute for 10 min. 5. The tube that cools fastest, emits infrared energy quickest.	• A common error in this experiment is not having the boiling tubes at the same temperature at the start – a hotter tube will cool quicker initially, which can affect results. • Evaporation from the surface of the water can cause cooling too, which will affect the results. To minimise this, block the top of each tube with a bung or a plug of cotton wool.
Variables	**Hazards and Risks**
• The independent variable is the colour of the boiling tube. • The dependent variable is the temperature. • Control variables include volume of water, start temperature and environmental conditions.	• The main hazard is being burned when pouring the hot water and when handling the hot tubes. Using a test tube rack to hold the tubes minimises the need to touch the tubes and means hands can be kept clear when pouring the water into them.

HT Radio Signals

- Radio waves can be caused by oscillations in electrical circuits, i.e. an alternating current (see pages 60–61).
- The frequency of the radio wave produced matches the frequency of the electrical oscillation. This is how a radio signal is produced.
- When radio waves are absorbed by a conductor they may create an alternating current with the same frequency as the radio wave, this is how the signal is received.
- When this oscillation is induced in an electrical circuit it creates an electrical signal that matches the wave.

Hazards of EM Waves

- Changes in atoms and the nuclei of atoms can result in EM waves being generated or absorbed over a wide frequency range:
    - Electrons moving between energy levels as a result of heat or electrical excitation can generate waves, e.g. infrared waves, visible light, ultraviolet waves and X-rays.
    - Changes in the nucleus of an atom can generate waves, i.e. an unstable nucleus can give out excess energy as gamma rays.
- Ultraviolet waves, X-rays and gamma rays carry enough energy to have hazardous effects on the human body:
    - Ultraviolet waves can cause the skin to age prematurely and increase the risk of skin cancer.
    - X-rays and gamma rays are ionising radiation – they can damage cells by ionising atoms and, if absorbed by the nucleus of the cell, can cause gene mutations and cancer.

> **Key Point**
>
> The risk of damage from EM waves depends on the type of radiation and the amount of exposure.
>
> Radiation dose is a measure of harm based on these two factors. It is measured in Sieverts (Sv).

Radiation Damage

Ionising radiation source

The irradiated cell may...

...suffer no damage. | ...mutate, which can lead to cancer. | ...die, leading to burns, sickness and even death.

> **Key Words**
>
> electromagnetic (EM) waves
> microwaves
> infrared
> ultraviolet
> X-rays
> gamma
> ionising

> **Quick Test**
>
> 1. List the types of wave in the electromagnetic spectrum in order, from long wavelength to short wavelength.
> 2. Give **one** example of how electromagnetic waves transfer energy.
> 3. HT What property of microwaves makes them suitable for cooking?

Lenses

You must be able to:

- Describe what is meant by a real and virtual image
- Construct ray diagrams for both concave and convex lenses
- Calculate the magnification produced by a lens.

Lenses

- A lens forms an image by refracting light.
- There are two main types of lens: convex and concave.
- A convex lens is wider in the middle than at the edges:
  - Parallel rays of light entering a convex lens are brought to a focus at the principal focus or focal point.
  - Because parallel rays of light entering a convex lens converge (come together), they are sometimes called converging lenses.
- The distance from the lens to the principal focus is the focal length.
- A concave lens is wider at the edges than it is in the middle:
  - Parallel rays of light entering a concave lens spread out.
  - This makes the rays appear to have come from the principal focus on the same side of the lens that they originated.
  - Because parallel rays of light entering a concave lens diverge, they are sometimes called diverging lenses.

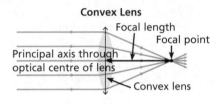

Convex Lens

Focal length
Focal point
Principal axis through optical centre of lens
Convex lens

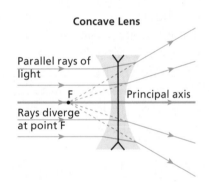

Concave Lens

Parallel rays of light
F
Principal axis
Rays diverge at point F

Images and Magnification

- Convex lenses can produce real or virtual images.
- Concave lenses only produce virtual images.
- A real image is on the opposite side of the lens to the object and can be projected onto a screen.
- A virtual image is on the same side as the object and can only be seen by looking through the lens.

Ray Diagrams

- When drawing ray diagrams:
  - Draw the principal axis. This is a horizontal line that runs straight through the lens.
  - Use the correct lens symbols (do not draw a picture of the lens).
  - Mark the principal foci on either side of the lens by drawing a dot on the principal axis and labelling it F.
  - Mark the position of the object as an arrow standing on the principal axis.
  - Once the diagram is laid out, it is time to draw the light rays.
- Magnification is the ratio of image height to object height, i.e. a magnification of 2 means the image is twice the size of the object.
- As magnification is a ratio, it has no units.

> ### Key Point
>
> The correct lens symbols are:
>
> convex lens =
>
>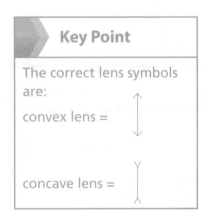
>
> concave lens =

$$\text{magnification} = \frac{\text{image height}}{\text{object height}}$$

Convex Lens	Concave Lens
1 Draw a horizontal line that runs straight from the top of the object to the lens.	**1** Draw a horizontal line that runs from the top of the object to the lens.
2 Draw a second line from the point where the first line meets the lens down through the principal focus on the far side of the lens.	**2** Draw a dotted line from the principal focus on the same side as the object to the point where the first line meets the lens. Continue this second line through the lens as a solid line.
3 Draw a third line from the top of the object, running diagonally through the middle of the lens and out the other side.	**3** Draw a third line from the top of the object running diagonally through the middle of the lens.
4 If the lines meet on the opposite side of the lens, it is a real image. Draw the image as a vertical arrow connecting the principal axis to the point where the lines cross.	**4** Where the dotted line meets the third line marks the top of the image, this will be a virtual image and will be the right way up. Draw a vertical arrow on this image from the principal axis to where the lines meet.

Distant object (beyond '2F')

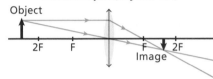

If the ray lines are diverging, it is a virtual image:
- Trace the lines backwards, as dotted lines, until they meet.
- Draw the image at this point, as a vertical arrow from the principal axis to the point where the lines meet.

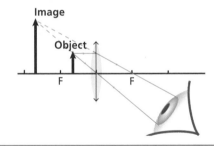

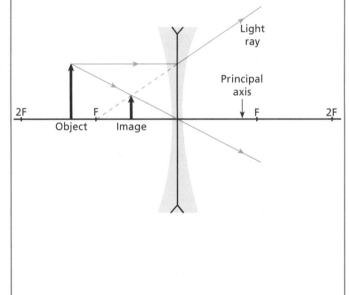

Quick Test

1. How is a real image different from a virtual image?
2. Which type of lens is wider in the middle than at the edges?
3. A convex lens provides a magnification of 0.75. The object is 4cm high. What will the height of the magnified image be?

Light and Black Body Radiation

You must be able to:

- Explain what is meant by 'visible light' and 'a perfect black body'
- Explain why an object appears a particular colour
- Explain the effect of coloured filters
- Explain how temperature affects emitted radiation
- Explain how the temperature of the Earth is determined by emitted and absorbed radiation.

Visible Light

- 'Visible light' describes electromagnetic waves that can be detected by the human eye.
- When light is incident on (arrives at and hits) an object, it can be absorbed, reflected or transmitted.
- Reflection by a smooth surface in a single direction (e.g. by a mirror) is called specular reflection.
- Reflection from a rough surface, where the light is scattered, is called diffuse reflection.
- All objects are either:
  - transparent – they transmit light coherently (the light rays do *not* get jumbled up) so objects on the other side can be seen clearly
  - translucent – they transmit light, but the rays are scattered so objects cannot be seen clearly through them, e.g. frosted glass
  - opaque – they either reflect or absorb all light incident on them, so no light passes through them.

Colour

- Each colour within the visible spectrum has its own narrow band of wavelength and frequency.
- When an opaque object appears coloured:
  - it is reflecting light of that particular wavelength
  - it is absorbing other wavelengths.
- If all wavelengths are reflected equally, the object appears white.
- If all wavelengths are absorbed, the object appears black.
- Coloured filters work by absorbing some wavelengths and not others.
- The wavelengths that are transmitted control what colour the filter allows to pass through.
- When observing coloured objects through coloured filters a variety of different effects can be seen.
- If the filter is the same colour as the object, the object will appear its true colour.
- If the object is a different colour to the filter then a few things could happen:
  - A red and blue striped object seen through a red filter will appear red and black. This is because the filter will allow the red light through, but not the blue light.
  - The same object seen through a green filter will appear completely black. This is because the filter will not allow red or blue to pass through it.

> ### Key Point
>
> A common mistake is to think that a filter produces a colour mixing effect when viewing coloured objects. This is not the case – a coloured filter does not change the colour – it filters out some colours while letting others through.

Infrared Radiation

- All bodies (objects) **emit** and **absorb** infrared radiation.
- The rate at which an object emits radiation depends on the nature of the surface and on its temperature – the hotter the body, the faster it emits infrared radiation.
- A perfect **black body**:
  - absorbs all of the infrared radiation incident on it
  - does not reflect or transmit any infrared radiation.
- Since a good absorber is also a good emitter, a perfect black body is also the best possible emitter.

Black Bodies

- The temperature of an object (or body) determines:
  - the rate at which it emits radiation
  - the wavelength of the radiation it emits.
- As the temperature increases the amount of radiation an object emits at all wavelengths increases, but the intensity of shorter wavelengths increases faster.
- As an object is heated, it first glows red hot. As it gets hotter, it emits even shorter wavelengths and it becomes whiter.

HT The temperature of an object is related to the balance between radiation absorbed and radiation emitted.

HT The ground on a sunny day will increase in temperature when the Sun comes out.

HT This is because it absorbs radiation from the Sun faster than it emits radiation.

HT As the ground gets warmer, the rate at which it emits radiation will increase.

HT Eventually, the rate of emission is equal to the rate of absorption, and the ground will then be at constant temperature.

HT This effect applies to many other situations, e.g. a radiator cooling down, a house, an object in front of a radiant heater and the planet Earth itself.

HT The temperature of the Earth depends on many things:
  - how much energy it receives from the Sun
  - how much energy is reflected back into space
  - how much energy it emits into space.

HT The Earth's atmosphere also affects how much of the radiation emitted from the surface escapes into space.

Blackbody Radiation Curves

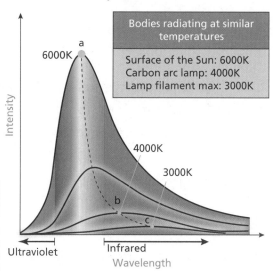

Bodies radiating at similar temperatures

Surface of the Sun: 6000K
Carbon arc lamp: 4000K
Lamp filament max: 3000K

Quick Test

1. How is a transparent object different from a translucent one?
2. Explain why a green object appears black if viewed through a red filter.
3. The coals in a barbecue glow red. Burning magnesium is bright white. What does the colour tell you about the temperature of these two things?

Review Questions

Forces – An Introduction

1 HT A cyclist comes to a downward slope in the road and stops pedalling.

The free body diagram in **Figure 1** shows the different forces acting on a bicycle as it travels down the hill.

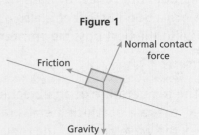

Figure 1

a) Draw a single arrow on **Figure 1** to show the total effect of all these forces. [1]

b) From the resultant force arrow you drew for part **a)**, determine if the bicycle is speeding up, slowing down or staying at constant speed. [1]

2 Explain the difference between a contact force and a non-contact force. [2]

3 Complete the sentences.

There are two different types of force, contact and non-contact forces. .... ,

.... and electrostatic forces are examples of .... forces. Friction and

.... are .... forces that oppose motion. [5]

4 The gravitational field strength on the Moon is 1.6N/kg and the gravitational field strength of Earth is 10N/kg.

a) Calculate the weight of a 70kg astronaut on **i)** the Moon and **ii)** on Earth. [2]

b) Explain why an astronaut can jump higher on the Moon than on Earth. [2]

5 HT Use a vector diagram to show an object travelling with a driving force of 20N and a frictional force of 5N.

Your diagram should show:
- The driving force
- The frictional force
- The resultant force
- The size of the resultant force. [4]

6 HT An arrow in flight experiences air resistance of 0.5N and a gravitational force of 2N.

Show this on a scale vector diagram and add an arrow to indicate the resultant force. [3]

Total Marks .... / 20

Forces in Action

1 Write down the equation that shows the relationship between work done, force and distance. [1]

2 In a tall building, the height between floors is 3.5m.
The lift car that carries people between floors weighs 1200N.

Calculate the work done by the engines when the lift car is raised up five floors. [3]

3 **Figure 1** shows a flight of stairs.

Calculate the work done by a 40kg child climbing the stairs (g = 10N/kg). [3]

4 Explain why at least two forces are needed to stretch a spring. [2]

5 Using springs as an example, explain what is meant by the 'limit of proportionality'. [2]

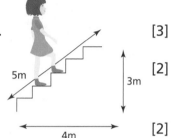

Figure 1

6 A fisherman fixes a 4m fishing rod at the base creating a pivot.
He holds the rod 1m from the pivot.
A fish is caught and pulls away with a force of 12N.

Calculate the force the fisherman must apply to hold the rod still. [4]

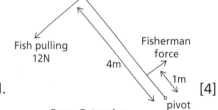

7 A student carries out an experiment to investigate the extension of a rubber band.
The results are plotted in a graph of force over extension.
The line produced is curved and gets steeper as the force increases.

What conclusions can be drawn from the graph? [3]

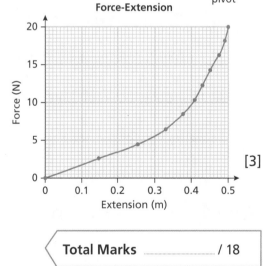

Total Marks _____ / 18

Pressure and Pressure Differences

1 At sea level, the atmospheric pressure is 100 000N/m².

a) Calculate the weight of air that would act on a house roof with an area of 10m². [2]

b) Explain why this weight does not cause the roof to cave in. [3]

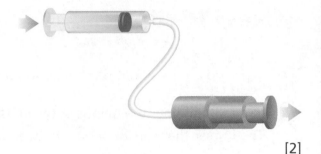

2 In a hydraulic system, the cross-sectional area of the left-hand piston is 0.1cm². The area of the right-hand piston is 0.5cm².

If the left-hand piston is pushed down with a force of 200N, what force will be transferred to the right-hand piston? [2]

3 HT For this question, you will need to refer to the Physics Equations on page 140.

Gravitational field strength is 10N/kg.
The density of mercury is 13 500kg/m³.

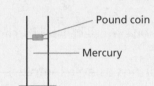
Pound coin
Mercury

In a laboratory, a column of mercury is poured into a measuring cylinder with a cross-sectional area of 3cm².

a) Calculate the pressure at the bottom of a 10cm tall column of mercury. [3]

b) Calculate the weight of mercury that acts on the bottom of the measuring cylinder. [3]

c) When a pound coin is dropped into the mercury, it floats on the surface.

What conclusion can be drawn about the density of the pound coin? [1]

d) The pound coin has a weight of 0.1N.

Work out the weight of mercury that is displaced by the coin. [1]

4 At the top of a mountain the air pressure is lower than at sea level.

Explain why this is. [3]

> Total Marks .... / 18

Forces and Motion

1 Describe the difference between speed and velocity. [2]

2 Write down the formula that shows the relationship between speed, distance and time. [1]

3 A hiker travels south for three miles, west for two miles and then north for one mile.

a) What is the total distance travelled by the hiker? [1]

b) HT Use a scale vector diagram to show the final displacement of the hiker. [2]

4 A sound wave travelling in the ocean takes 3 seconds to travel 4.5km.

Calculate the speed of sound in water in metres per second (m/s). [2]

5 On a distance–time graph, what do the following features represent?

a) A straight line with a very steep gradient. [1]

b) A horizontal line. [1]

c) A curved line. [1]

6 Sketch a distance–time graph to show the motion of an object that:
- Accelerates gradually to a constant speed of 1m/s.
- Remains at this speed for 2 seconds.
- Decelerates gradually to come to rest having travelled 8 metres in total. [4]

7 HT Using an example, explain how an object can travel at constant speed but with a changing velocity. [3]

8 HT What name is given to the tendency of objects at rest to remain at rest and for moving objects to keep moving? [1]

9 Use the force arrows to work out the resultant force and the motion of each car. [4]

a)
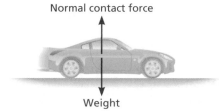
Normal contact force

Weight

b)
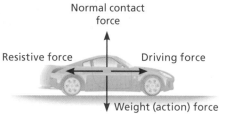
Normal contact force

Resistive force   Driving force

Weight (action) force

Total Marks _____ / 23

Forces and Acceleration

1 Write down whether each of the following statements is **true** or **false**.

a) If the resultant force on an object is doubled, its acceleration will double. [1]

b) If the force on an object is constant and the mass of the object increases, the acceleration will also increase. [1]

c) Newton's second law states that for every force there is an equal and opposite reaction force. [1]

Review Questions

2 A car travelling at 20m/s decelerates at a constant rate of 2m/s².

Use the formula $v^2 - u^2 = 2as$ to calculate how far the car travels before coming to rest. [3]

3 State what the following features would represent on a velocity–time graph.

a) A straight line with a positive gradient. [1]

b) A straight line with a negative gradient. [1]

c) A horizontal line above the x-axis. [1]

d) A horizontal line below the x-axis. [1]

e) The area under the graph. [1]

Total Marks .... / 11

Terminal Velocity and Momentum

1 The velocity–time graph in **Figure 1** shows how the velocity of a skydiver changes, from the moment they jump out of the aircraft, past the point where the parachute is deployed, to the point where they land.

Which number on the graph represents:

Figure 1

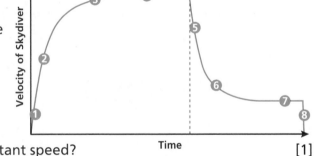

a) A point where the skydiver is travelling at constant speed? [1]

b) A point just after the parachute has been deployed? [1]

c) A point where the force from gravity is much larger than the air resistance? [1]

d) A point at which the vertical forces are balanced? [1]

2 Use Newton's third law and the idea of equal and opposite forces to explain why a hovercraft moves forward when the propeller spins. [3]

3 HT Write down the formula that links momentum, mass and velocity. [1]

4 HT A 240kg cannon fires a 1kg cannonball with a velocity of 120m/s.

Work out the velocity at which the cannon recoils backwards when it is fired. [4]

5 HT On a snooker table, the cue ball is travelling at 4m/s when it strikes the yellow ball. The cue ball continues at $\frac{1}{4}$ of its original speed.

If both balls are the same mass, what speed does the yellow ball travel at? Assume the yellow ball travels in the same direction as the cue ball. [2]

Total Marks _____ / 14

Stopping and Braking

1 Give **three** factors that would have a negative effect on the braking distance of a vehicle. [3]

2 HT Two students, Lisa and Jack, are discussing the braking of vehicles. Jack thinks that if the braking force is the same but a car is going twice as fast, it will take twice the distance to stop.

Is he correct? You must explain your answer. [3]

3 Many long, downhill roads have an escape lane at the bottom for a driver to turn into if their brakes fail.

Explain why brakes are more likely to fail while travelling down a long, steep hill. [4]

4 Explain what is meant by 'reaction time' and how this applies to thinking and stopping distances. [3]

5 A car travelling at 12m/s has a braking distance of 14m.

If the driver's reaction time is 0.5 seconds, what is the total stopping distance? [3]

6 State whether each of these statements about stopping distances is **true** or **false**.

a) The thinking distance is affected by the speed of the car. [1]

b) Poor road and weather conditions will increase the braking distance. [1]

c) The higher the speed, the greater the overall stopping distance. [1]

d) Fatigue and tiredness will increase the thinking distance. [1]

Total Marks _____ / 20

Energy Stores and Transfers

1   The useful energy output from a petrol engine is kinetic energy. However, the engine wastes more energy as heat than it produces as kinetic energy. This and other transfers involved are shown on the Sankey diagram in **Figure 1**.

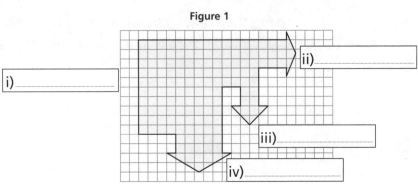

Figure 1

i) ....

ii) ....

iii) ....

iv) ....

a) Label the Sankey diagram in **Figure 1** using words from the box.   [4]

| Chemical energy | Kinetic energy | Heat energy | Sound energy |

b) If the total input energy is 1000kJ, how much energy is converted to kinetic energy?   [2]

2   A skateboarder of mass 60kg is practising on a half-pipe.
She starts at the top, skates down one side and back up the other side.

a) Describe the energy transfers involved.   [6]

b) The maximum speed the skateboarder reaches is 4m/s.

Calculate her kinetic energy at this point.   [3]

c) The skater starts at a height of 2m.

Calculate how much gravitational energy she has at this point (g = 10N/kg).   [3]

3   A large spring is stretched by 10cm and when released returns to its original shape.
The spring constant is 100N/m.
elastic potential energy = 0.5 × spring constant × (extension)^2

Calculate the amount of energy that has been stored in the spring.   [3]

Total Marks .... / 21

Energy Transfers and Resources

1   Coal is a non-renewable energy resource. What is meant by 'non-renewable'?   [1]

2 Explain the different factors that determine the amount of heat lost from a building. [3]

Total Marks _____ / 4

Waves and Wave Properties

1 Give **one** example to show that waves do not transfer matter. [2]

2 Complete the sentences to describe the process by which energy is transferred by a wave.
Use answers from the box.

oscillate	fixed	particles	energy

The _____ that make up a wave _____ about a _____ point.

When doing this, they pass _____ onto adjacent particles and start them oscillating. [4]

3 **Figure 1** shows a wave.

Figure 1

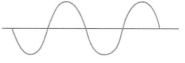

a) Add labels to **Figure 1** to show the wavelength and amplitude. [2]

b) How many complete waves are shown in **Figure 1**? [1]

c) The waves shown have a period of 2 seconds.

How long would the waves pictured take to pass a fixed point? [1]

4 A sound wave travels at a speed of 330m/s in air.

If the frequency of the wave is doubled, what effect would this have on the wavelength of the wave? [1]

5 A low frequency sound wave has a time period of 0.05s and travels at a speed 330m/s.

$$period = \frac{1}{frequency}$$

a) Calculate the frequency of the wave. [2]

b) Use the wave equation to calculate the wavelength of the wave. [2]

6 Describe a method that can be used to measure the speed of sound in air. [4]

Total Marks _____ / 19

Practice Questions

Reflection, Refraction and Sound

1 At a blind road junction, a mirror is put up to allow the drivers on the minor road to see if cars are approaching from the right.

On **Figure 1** draw:
- Where the mirror should be placed.
- The rays of light reflected from the mirror that would allow the driver in the car on the minor road to see the car approaching on the main road. [4]

Figure 1

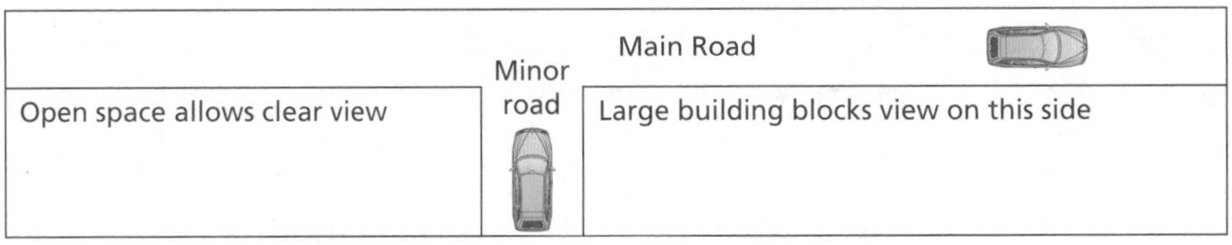

	Main Road	
Open space allows clear view	Minor road	Large building blocks view on this side

2 **Figure 2** shows a ray of light entering a glass block.

Figure 2

Complete **Figure 2** to show:
- The path of the light ray as it passes through the block.
- The normal where the ray enters and exits the block. [4]

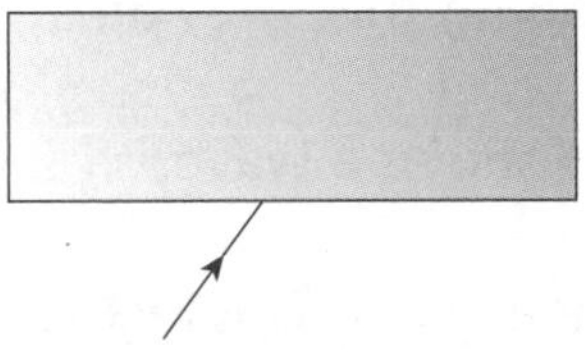

3 HT What happens to a wave as it passes from one medium to another?
Circle **one** answer.

Its frequency changes   **Nothing happens**   **Its speed changes** [1]

4 How does the speed of light in a vacuum differ from the speed of light in glass? [1]

5 a) What property of a sound wave affects its loudness? [1]

b) What property of a sound wave affects its pitch? [1]

Total Marks _____ / 12

HT Waves for Detection and Exploration

1 A factory produces containers for pressurised air.
An ultrasound scan is used to check if there are any faults in the containers and to make sure that the walls are the correct thickness.
A batch of containers is found to have walls that are much thinner than normal.

Explain how the ultrasound scan would have provided this information. [2]

2 A fishing ship uses echo sounding to locate a shoal of fish.
The ship sends an ultrasound pulse.
It detects a reflected pulse from the shoal 0.1 seconds afterwards.

Calculate how far away the shoal is from the ship (speed of sound in water = 1500m/s). [4]

3 a) Write down the **two** types of seismic waves. [2]

b) Use your knowledge of seismic waves to describe how they can provide evidence about the location of the earthquake that produced them. [2]

4 a) What is meant by a seismic wave 'shadow zone'? [1]

b) What do these shadow zones prove about the structure of the Earth? [2]

Total Marks .... / 13

The Electromagnetic Spectrum

1 a) Mobile phones send signals using electromagnetic waves.

Which type of electromagnetic wave is used? [1]

b) A scientist being interviewed for television says:

I think that using mobile phones too much increases the risk of developing a brain tumour.

Is this a conclusion, a fact or an opinion? [1]

c) A study on the link between mobile phones and brain tumours observed 420 000 adults over a 10-year period.
The study found no correlation between the amount of mobile phone use and brain tumours.

i) Which of the following conclusions is most accurate?
Tick **one** box.

Short to medium-term use by adults does not cause brain tumours. ☐

Mobile phones are completely safe to use. ☐

Children should not use mobile phones. ☐

It is not safe for adults to use mobile phones. ☐ [1]

ii) Would you expect mobile phone use to result in cancer?
Use your knowledge of electromagnetic waves and their hazards to explain your answer.   [3]

Total Marks _____ / 6

Lenses

1 Lenses can make use of refraction to bring light to a focus.

Describe the shape of a converging lens.   [2]

2 **Figure 1** shows a convex lens.

Figure 1

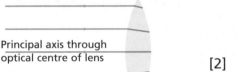

Copy the diagram and add ray lines and arrows to
Figure 1 to show the path of the light rays as they pass
through the lens and are brought to a focus.

Principal axis through
optical centre of lens

[2]

3 Give a definition for the 'focal length' of a lens.   [1]

4 Complete the sentences using words from the box.   [4]
You may use the words once, twice or not at all.

| parallel | principal axis | focal point | focus | distant | near |

Light rays from _____ objects are effectively _____ . When a convex lens

focuses light from these objects, the image will be formed at the _____ . If the object

is in line with centre of the lens, the image will be formed on the _____ .

5 Complete the ray diagram in **Figure 2** to
show the image formed by a convex lens when
it is being used as a magnifying glass and
produces a virtual image.

Figure 2

Object

F   F

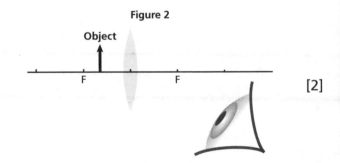

[2]

Total Marks _____ / 11

Light and Black Body Radiation

1 Name and explain the **two** different types of reflection that can occur when light is incident on a surface. [4]

2 Draw **one** line from each term to the way in which it makes light behave.

Transparent		Light passes through and remains coherent so objects can be seen clearly.
Translucent		No light passes through.
Opaque		Light rays pass through but become 'jumbled' so that objects are obscured.

[2]

3 A man owns a blue car.

He notices that the car looks a much darker shade of blue when parked under an orange street light compared to when it is parked under a white street light.

On very dark nights, with no Moon and few other light sources, it looks almost black under the orange light.

Explain why this is. [4]

4 **a)** How does the rate at which an object emits electromagnetic radiation depend on its temperature? [1]

b) Describe what happens to the appearance of an object as it increases in temperature. [3]

5 HT This question is about absorbing and emitting infrared radiation.

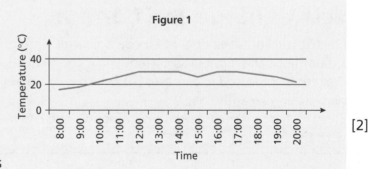

The graph in **Figure 1** shows the temperature recorded in a car parked in the sun during one day.

Figure 1

a) Explain why the temperature increases initially. [2]

b) Explain why the temperature levels off at a maximum of 30°C. [2]

c) Suggest what might have happened at around 3pm to cause the temperature inside the car to fall. [2]

Total Marks .... / 20

An Introduction to Electricity

You must be able to:

- Draw and interpret circuit diagrams
- Calculate the charge that flows in a circuit
- Relate current, resistance and potential difference
- Explain how to investigate factors that affect the resistance of an electrical component.

Standard Circuit Symbols

- In diagrams of electrical circuits:
  - standard circuit symbols are used to represent the components
  - wires should be drawn as straight lines using a ruler.
- You need to know all of the circuit symbols in the table below:

Component	Symbol	Component	Symbol
Switch (open)		LED (light emitting diode)	
Switch (closed)		Bulb / lamp	
Cell		Fuse	
Battery		Voltmeter	
Diode		Ammeter	
Resistor		Thermistor	
Variable resistor		LDR (light dependent resistor)	

Electric Charge and Current

- Electric current is the flow of electrical charge – the greater the rate of flow, the higher the current.
- Current is measured in amperes (A), which is often abbreviated to amps, using an ammeter.
- Electric charge is measured in coulombs (C) and can be calculated with the equation:

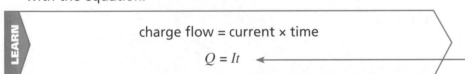

$$\text{charge flow} = \text{current} \times \text{time}$$
$$Q = It$$

- As the current in a single, closed loop of a circuit has nowhere else to go (i.e. no branches to travel down), the current is the same at all points in the loop (see page 69).

> **Key Point**
>
> An ammeter is connected in series. A voltmeter is connected in parallel to the component.

Charge flow (Q) is measured in coulombs (C).
Current (I) is measured in amps (A).
Time (t) is measured in seconds (s).

Resistance and Potential Difference

- The resistance of a component is the measure of how it resists the flow of charge.
- The higher the resistance:
  - the more difficult it is for charge to flow
  - the lower the current.
- Resistance is measured in ohms (Ω).
- Potential difference (or voltage) tells us the difference in electrical potential from one point in a circuit to another.
- Potential difference can be thought of as electrical push.
- The bigger the potential difference across a component:
  - the greater the flow of charge through the component
  - the bigger the current.
- Potential difference is measured in volts (V) using a voltmeter.
- Potential difference, current and resistance are linked by the equation:

LEARN

> potential difference = current × resistance
> $$V = IR$$

Potential difference (V) is measured in volts (V).
Current (I) is measured in amps (A).
Resistance (R) is measured in ohms (Ω).

REQUIRED PRACTICAL

Investigate the factors that affect the resistance of an electrical component.

Sample Method	**Considerations, Mistakes and Errors**
This example looks at how length affects the resistance of a wire: 1. Set up the standard test circuit as shown. 2. Pre-test the circuit and adjust the supply voltage to ensure that there is a measurable difference in readings taken at the shortest and longest lengths. 3. Record the voltage and current at a range of lengths, using crocodile clips to grip the wire at different points. 4. Use the variable resistor to keep the current through the wire the same at each length. 5. Use the voltage and current measurements to calculate the resistance.	• Adjusting the supply voltage to ensure as wide a range of results as possible is important, as measurements could be limited by the precision of the measuring equipment. • The range of measurements to be tested should always include at least five measurements at reasonable intervals. This allows for patterns to be seen without missing what happens in between, but also without taking large numbers of unnecessary measurements.
Variables	**Hazards and Risks**

• The independent variable is the length of the wire.
• The dependent variable is the voltage.
• The control variable is the current (which is kept the same, because if it was too high it would cause the wire to get hot and change its resistance).
	• Current flowing through the wire can cause it to get very hot.
	• To avoid being burned by the wire:
	  – a low supply voltage should be used, such as the cell in the diagram
	  – adjust the variable resistor to keep the current low.

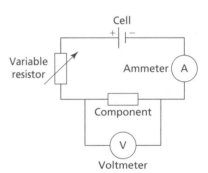

Cell

Variable resistor

Ammeter (A)

Component

Voltmeter (V)

Quick Test

1. A 600C charge flows in a circuit in 1 minute. What current is flowing?
2. Draw a standard test circuit for an investigation to measure the resistance of a diode.
3. A bulb has a 10V potential difference across it. A 2A current flows. What is the resistance of the bulb?

Circuits and Resistance

You must be able to:

- Interpret potential difference–current graphs
- Describe the shape of potential difference–current graphs for various components
- Use circuit diagrams to construct circuits.

Resistors and Other Components

- Potential difference–current graphs (V–I graphs) are used to show the relationship between the potential difference (voltage) and current for any component.
- A straight line through the origin indicates that the voltage and current are directly proportional, i.e. the resistance is constant.
- A steep gradient indicates low resistance, as a large current will flow for a small potential difference.
- A shallow gradient indicates a high resistance, as a large potential difference is needed to produce a small current.
- For some resistors the value of R is not constant but changes as the current changes; this results in a non-linear graph.

> **Key Point**
>
> A V–I graph shows the relationship between voltage and current. It therefore can be used to determine the resistance.

REQUIRED PRACTICAL	
Investigate the V–I characteristics of a filament lamp, a diode and a resistor at constant temperature.	
Sample Method 1. Set up the standard test circuit as shown. 2. Use the variable resistor to adjust the potential difference across the test component. 3. Measure the voltage and current for a range of voltage values. 4. Repeat the experiment at least three times to be able to calculate a mean. 5. Repeat for the other components to be tested.	**Considerations, Mistakes and Errors** • Before taking measurements, check the voltage and current with the supply turned off. This will allow zero errors to be identified. • A common error is simply reading the supply voltage as the voltage across the component. At low component resistances, the wires will take a sizeable share of this voltage, resulting in a lower voltage across the component. This is why a voltmeter is used to measure the voltage across the component.
Variables • The independent variable is the potential difference across the component (set by the variable resistor). • The dependent variable is the current through the component, measured by the ammeter.	**Hazards and Risks** • The main risk is that the filament lamp will get hotter as the current increases and could cause burns. If it overheats, the bulb will 'blow' and must be allowed to cool down before attempting to unscrew and replace it.

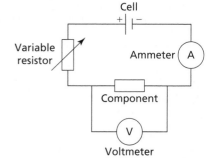

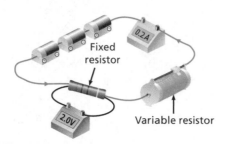

Resistors
- An **ohmic conductor** is a resistor in which the current is directly proportional to the potential difference at a constant temperature.
- This means that the resistance remains constant as the current changes.
- It is indicated by a linear (straight line) graph.

Filament Lamps
- As the current through a filament lamp increases, its temperature increases.
- This causes the resistance to increase as the current increases.
- It is indicated by a curved graph.

Diodes
- The current through a **diode** will only flow in one direction.
- The diode has a very high resistance in the reverse direction.
- This is indicated by a horizontal line along the *x*-axis, which shows that no current flows.

Thermistors
- The resistance of a thermistor decreases as the temperature increases.
- This makes them useful in circuits where temperature control or response is required.
- For example, a thermistor could be used in a circuit for a thermostat that turns a heater off at a particular temperature or an indicator light that turns on when a system is overheating.

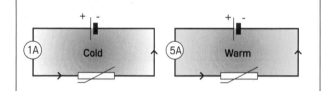

Light Dependent Resistors (LDRs)
- The resistance of an LDR decreases as light intensity increases.
- This makes them useful where automatic light control or detection is needed, e.g. in dusk till dawn garden lights / street lights and in cameras / phones to determine if a flash is needed.

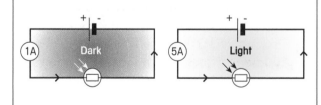

Quick Test

1. A *V–I* graph is plotted for a component. When the potential difference is negative, no current flows. What component has been tested?
2. A *V–I* graph is steep for high temperatures and shallow for low temperatures. What component does it represent?

Key Words

ohmic conductor
diode

Circuits and Power

You must be able to:

- Explain the difference between series and parallel circuits
- Explain the effect of adding resistors in series and parallel.
- Explain what is meant by 'power' using examples
- Explain how the power transfer in any circuit device is related to the potential difference across it and the current through it, and to the energy changes over time.

Series and Parallel Circuits

- Electrical components can be connected in series or parallel.
- Some electrical circuits contain series and parallel parts.

Series Circuits	Parallel Circuits
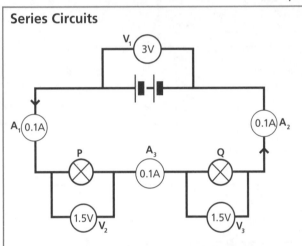	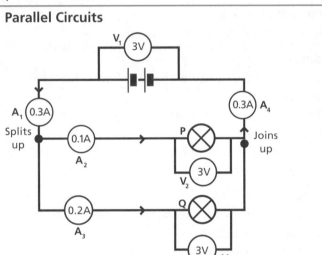
• There is the same current through each component. • The total potential difference of the power supply is shared between the components. • The total resistance of two components is the sum of the resistance of each component. This is because the current has to travel through each component in turn. • Adding resistors in series increases the total resistance (*R*) in ohms (Ω): $$R_{total} = R_1 + R_2$$	• The potential difference across each component is the same. • The total current drawn from the power supply is the sum of the currents through the separate components. • The total resistance of two resistors is less than the resistance of the smallest individual resistor. This is because, in parallel, there are more paths for the current to take – it can take one or the other, allowing it to flow more easily. • Adding resistors in parallel reduces the total resistance.

- You need to be able to calculate the currents, potential differences and resistances in d.c. series circuits.

> A circuit containing two resistors in series has a 12V supply.
> R_1 is a 4Ω resistor and has a voltage of 8V across it.
>
> **a)** Work out the voltage across R_2.
>
> $V_{total} = V_1 + V_2$ ← Remember, in a series circuit, the power supply voltage is shared.
> $12 = 8 + V_2$ ← Substitute in the given values.
> $V_2 = 12 - 8 = 4V$ ← Rearrange to find the voltage across the second resistor.

b) Calculate the current that flows in the circuit.

$V = IR$

$I = \dfrac{V}{R} = \dfrac{8}{4} = 2A$ ◄ The current flow is the same through every component, so it can be calculated using the known voltage and resistance of R_1.

c) Work out the resistance of R_2.

$V = IR$ ◄ Use the values for voltage and current calculated in parts **a)** and **b)**.

$R = \dfrac{V}{I} = \dfrac{4}{2} = 2\Omega$

You could also use equivalent resistance to answer part **c)**:

From the 12V supply, R_1 took 8V and R_2 took 4V. R_1 must be twice as difficult for current to flow through, i.e. it must have twice the resistance. Therefore, $R_2 = 2\Omega$

Power in Circuits

- The **power** of a device depends on the potential difference across it and the current flowing through it.
- A device with a higher potential difference or current will use more energy per second than one with a lower potential difference or current, i.e. it will be more powerful.

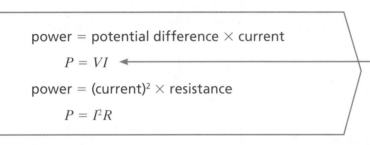

power = potential difference × current

$P = VI$

power = (current)² × resistance

$P = I^2R$

Power (P) is measured in watts (W). Potential difference (V) is measured in volts (V). Current (I) is measured in amps (A). Resistance (R) is measured in ohms (Ω).

The heating element in a kettle produces an output power of 2300W when a potential difference of 230V is applied across it.

a) Calculate the current flowing across the element.

$P = VI$

$I = \dfrac{P}{V}$

$I = \dfrac{2300}{230} = 10A$

b) Work out the resistance of the element.

$P = I^2R$

$R = \dfrac{P}{I^2}$

$R = \dfrac{2300}{10^2} = \dfrac{2300}{100} = 23\Omega$

You could also use $V = IR$ here:

$R = \dfrac{V}{I} = \dfrac{230}{10} = 23\Omega$

This is a useful way to double check your answers.

Quick Test

1. A series circuit with two resistors has a 6V supply. R_1 has a 5V potential difference across it. What can be said about the resistance of R_2 and what voltage will it have across it?
2. A torch bulb has a resistance of 6Ω. A current of 2A flows through it. Calculate the total power used by the bulb.
3. An oven has a 230V supply and a 30A current. Work out the power of the oven.

Key Words

series
parallel
power

Domestic Uses of Electricity

You must be able to:

- Explain the difference between direct and alternating current
- Describe a three-core cable
- Explain why a live wire may be dangerous even when a switch in the circuit is open
- Calculate the power of a device
- HT Understand how efficiency can be increased.

Direct and Alternating Current

- A **direct current (d.c.)** supply:
  - has a potential difference that is always positive or always negative – the current direction is constant
  - is the type of current that is supplied by cells and batteries.
- An **alternating current (a.c.)** supply:
  - has a potential difference that alternates from positive to negative – the current direction alternates
  - is the type of current used in mains electricity.

Direct Current

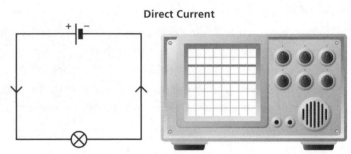

Alternating Current

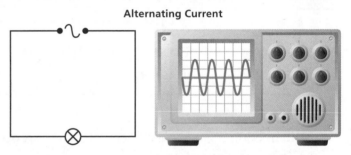

Mains Electricity

- Mains electricity in the UK is 230V and changes direction 50 times a second, i.e. it has a frequency of 50Hz.
- The mains supply uses three-core cable, i.e. the cable contains three wires.
- Each wire carries a different **electrical potential** and is colour-coded:
  - live wire (brown) – 230V potential
  - neutral wire (blue) – at or close to the 0V earth potential
  - earth wire (green and yellow stripes) – 0V potential.
- During operation:
  - the potential difference causes current to flow through the live and the neutral wires
  - the live wire carries the alternating potential from the supply
  - the neutral wire completes the circuit
  - current will only flow in the earth wire if there is a fault connecting it to a non-zero potential.
- The earth wire is a safety wire, which stops the exterior of an appliance becoming live.

The Three-Pin Plug

- **Casing**
- **Fuse**
- **Earth wire** (green and yellow)
- **Neutral wire** (blue) – carries current away from appliance
- **Cable grip** – secures cable in the plug
- **Live wire** (brown) – carries current to the appliance
- **Cable**
- 5A

Dangers of Mains Electricity

- Mains electricity can be very dangerous – an electric shock from a mains supply can easily be fatal.
- Touching the live wire can create a large potential difference across the body and result in a large current flowing through the body.

- The live wire can be dangerous even if a switch in the circuit is open.
- For example, a television might be switched off (so no current flows), but still plugged in and switched on at the wall:
  - The live wire between the wall and the switch on the television is still at an alternating potential.
  - All it needs is a path for the electricity to flow through.
  - This path could be provided by a damaged cable exposing the live wire.
  - If someone then touches the live wire, creating a potential difference from the live to the earth and causing current to flow, they will get an electric shock.

Power and Efficiency

- **Power** is the rate at which energy is transferred or work is done:

LEARN

$$\text{power} = \frac{\text{energy transferred}}{\text{time}} \quad \text{or} \quad \text{power} = \frac{\text{work done}}{\text{time}}$$

$$P = \frac{E}{t} \qquad\qquad P = \frac{W}{t}$$

Power (P) is measured in watts (W).
Energy transferred (E) is measured in joules (J).
Work done (W) is measured in joules (J).
Time (t) is measured in seconds (s).

- An energy transfer of 1J per second is equal to 1W of power.
- You must be able to demonstrate the meaning of power by comparing two things that use the same amount of energy but over different times.
- For example, if two kettles are used to bring the same amount of water to the boil and one takes less time, it is because it has a higher power.
- In an energy transfer, **efficiency** is the ratio of useful energy out to total energy in:
  - An efficiency of 0.5 or 50% means that half the energy is useful, but half is wasted.
  - An efficiency of 0.75 or 75% means that three-quarters of the energy is useful, but a quarter is wasted.

LEARN

$$\text{efficiency} = \frac{\text{useful energy transfer}}{\text{total energy transfer}} \quad \text{or}$$

$$\text{efficiency} = \frac{\text{useful power output}}{\text{total power input}}$$

HT To increase the efficiency of an energy transfer, the amount of wasted energy needs to be reduced.

Quick Test

1. Describe the difference between alternating current and direct current.
2. What is the voltage of the UK mains electricity supply?
3. A 2.5V bulb has a current of 1.2A flowing through it. What is the power of the bulb and how much energy would it use in 10s?
4. How much energy is transferred by a 2kW kettle in 50s?

Key Words

direct current (d.c.)
alternating current (a.c.)
electrical potential
power
efficiency

Electrical Energy in Devices

You must be able to:

- Describe how different domestic appliances transfer energy
- Calculate the energy transferred by a device
- Describe the structure of the national grid
- Explain why the national grid is efficient.

Energy Transfers in Appliances

- Whenever a charge flows, it has to overcome the resistance of the circuit. This requires energy, therefore:
  - work is done when charge flows
  - the amount of work done depends on the amount of charge that flows and the potential difference.
- The amount of energy transferred can also be found from the power of the appliance and how long it is used for, e.g. a 20W bulb uses 20J of energy in every second.

LEARN

energy transferred = power × time

$$E = Pt$$

energy transferred = charge flow × potential difference

$$E = QV$$

> Energy transferred (E) is measured in joules (J).
> Power (P) is measured in watts (W).
> Time (t) is measured in seconds (s).
> Charge flow (Q) is measured in coulombs (C).
> Potential difference (V) is measured in volts (V).

A 2kW heater is on for one hour. How much energy does it use?

1 hour = 60 minutes × 60 seconds = 3600 seconds
2kW = 2 × 1000W = 2000W

> Start by converting the values into watts and seconds.

$E = Pt$

$\quad = 2000 × 3600$

$\quad = 7\,200\,000J$

> The given values are for power and time, so use the first equation.

> Substitute in the values.

- Electrical appliances are designed to cause energy transfers.
- The type and amount of energy transferred between stores depends on the appliance.

Electrical energy 2000J/s

Heat energy (for element) 160J/s (wasted)

Heat energy (to water) 1800J/s (useful)

Sound energy 40J/s (wasted)

> ## Key Point
>
> The energy transferred by an appliance depends on the power and the time it is on for.

The National Grid

- The national grid is a system of cables and transformers linking power stations to homes and businesses.

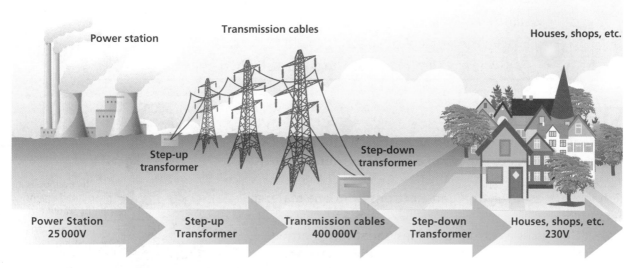

Power station

Transmission cables

Houses, shops, etc.

Step-up transformer

Step-down transformer

| Power Station 25 000V | Step-up Transformer | Transmission cables 400 000V | Step-down Transformer | Houses, shops, etc. 230V |

- Each component of the grid has a particular function.
- **Power station**:
  - The power station transfers the energy supply into electrical energy.
  - Using a smaller number of large power stations is more efficient than building many small, local power stations, because large stations can be made more efficient.
  - This is because most power plants use steam turbines, which are more efficient at higher steam temperatures, and the bigger the plant, the bigger the boiler, so the higher the steam temperature.
- **Step-up transformers**:
  - The transformers increase the potential difference from the power station to the transmission cables.
  - This reduces the current and, therefore, reduces the heating effect caused by current flowing in the transmission cables.
  - Reducing the heating effect reduces energy loss, so makes the transmission more efficient.
- **Transmission cables**:
  - Transmission cables transfer the electricity.
- **Step-down transformers**:
  - The transformers reduce the potential difference from the transmission cables to a much lower value for domestic use.

Quick Test

1. A bulb transfers 30J of energy in 10 seconds. During this time a charge of 12C is transferred. What is the voltage of the bulb?
2. Use a Sankey diagram to represent the energy transfers that take place in a washing machine that is 50% efficient.
3. A washing machine intensive cycle takes two hours. If the power is 500W, how much energy is transferred?
4. Give **two** ways in which the National Grid is designed to be an efficient way to transfer energy.

Key Words

National Grid
transformer

Static Electricity

You must be able to:

- Describe how static electricity is produced by rubbing surfaces
- Describe evidence that charged objects exert a non-contact force
- Explain how the transfer of electrons between objects can produce static electricity
- Explain the concept of an electric field
- Draw the electric field pattern around a charged object.

Static Charge

- When insulating materials are rubbed against each other they can become electrically charged, e.g. when a balloon is rubbed against a jumper.
- The friction moves negatively charged electrons from one material to another:
  - The object that gains electrons becomes negatively charged.
  - The object that loses electrons becomes positively charged.
- Because the materials are insulators, the charge remains on the object.
- This effect would not be seen with conductors – they conduct the charge to earth so it cannot build up.
- An object that has no conducting path to earth is referred to as an isolated object.

The balloon is rubbed against the jumper

Paper then clings to the balloon

Electrical Sparks

- As the charge on an isolated object increases, the potential difference between the object and earth increases.
- When the potential difference becomes high enough, a spark may jump across the gap, from the object to any earthed conductor near to it.
- This spark discharges the charged object and could be felt as an electric shock.
- It could also serve as a source of ignition, which can be very dangerous, e.g. if the spark occurs in a petrol station.
- Lightning is an example of spark caused when a charge builds up in clouds during a thunderstorm.

Charge on a Van de Graaff generator can discharge with a spark as the charge flows to earth.

Electrostatic Forces

- Electrostatic forces are non-contact forces.
- They can be forces of attraction or repulsion.
- If a charged object is brought near an uncharged object, it can attract it.

Perspex Rod Repels a Perspex Rod

Perspex Rod Attracts an Ebonite Rod

Perspex Rod Rubbed with Cloth

Ebonite Rod Rubbed with Fur

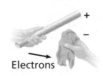
Electrons

Electrons

- This can be seen when a charged plastic ruler is brought near to small pieces of paper or close to water running from a tap.
- When both objects carry the same charge, they will repel.
- When the objects carry opposite charges, they will attract.

Electric Fields

- A charged object creates an electric field around itself.
- The electric field can be thought of as the area around a charged object that will affect other objects, e.g. like a magnetic field.
- The strength of the electric field at any point depends on two factors:
  - the distance from the object – the further away from the object, the weaker the field
  - the amount of charge – the higher the charge, the stronger the field.

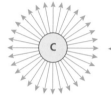

Tap

Charged plastic ruler

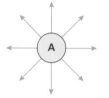 

The density of electric field lines around these three objects reveals that the quantity of charge on **C** is greater than that on **B** which is greater than that on **A**.

- If a second charged object is placed in the field, it will experience a force.
- The force gets stronger as the objects get closer together.
- You need to be able to draw the electric field pattern for an isolated charged sphere.
- A field diagram can help to explain electrostatic forces:
  - Where the field lines are close together, the field is stronger, so the force exerted on another object is stronger and the greater the chance a spark will occur.
  - The direction of the arrow indicates the direction that a positive charge will move in if placed in the field and explains why like charges repel and unlike charges attract.

Field lines

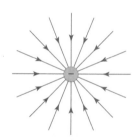

The Electric Field from an Isolated Positive Charge

The Electric Field from an Isolated Negative Charge

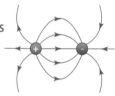 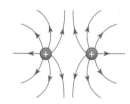

Quick Test

1. A child sliding down a plastic slide builds up a charge. They receive a shock at the bottom when they touch the metal legs of the slide. Explain why this would not happen with a metal slide.
2. Two identical balloons are charged by rubbing them on the same jumper. They are then suspended side by side. Describe what will happen.
3. How will the field lines drawn around a small negative charge be different from those around a large positive charge?
4. An object is moved further away from a charged object. Use the idea of electric fields to explain how this will affect the force on the object.

Key Point

All charged particles produce an electric field around themselves.

Key Words

isolated
spark
attraction
repulsion

Magnetism and Electromagnetism

You must be able to:

- Describe the difference between permanent and induced magnets
- Describe what happens when magnets interact
- Explain magnetic fields and magnetic field patterns
- Describe how the magnetic effect of a current can be demonstrated
- Explain why a solenoid can increase the magnetic effect of the current.

Magnetic Poles and Fields

- There are two types of magnetic pole: a north (seeking) pole and a south (seeking) pole.
- The poles of a magnet are the places where the magnetic forces are strongest.
- Unlike (opposite) poles attract – the north pole of a magnet will attract the south pole of another magnet.
- Like poles repel – a north pole will repel a north pole and a south pole will repel a south pole.
- The region around a magnet, where a force acts on another magnet or magnetic material (e.g. iron, steel or nickel cobalt), is called the magnetic field.
- The strength of the field depends on the distance from the magnet – it is strongest at the poles.
- Permanent magnets produce their own magnetic field.
- Induced magnets become a magnet when placed in a magnetic field. When removed from the field, they lose their magnetism quickly.
- The force between a permanent magnet and a magnetic material or an induced magnet is always one of attraction.
- The arrows on field lines always run from north to south and show the direction of the force that would act on a north pole placed at that point.
- The density of the field lines is called the flux density and indicates the strength of the field at that point – the closer together the lines, the higher the flux density.
- The higher the flux density, the stronger the field and the greater the force that would be felt by another magnet.

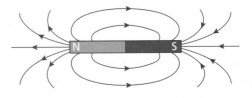

Plotting Fields

- A magnetic compass contains a small bar magnet.
- The compass needle aligns with the Earth's magnetic field and always points to the magnetic north.
- This provides evidence that the Earth's core must be magnetic.
- A magnetic compass can be used to plot the field around a bar magnet:
  1. Place the bar magnet on a piece of paper.
  2. Place the compass at one end of the magnet.
  3. On the paper, mark where the point of the compass needle is.
  4. Move the compass so the tail of the needle is at the point that has just been marked.

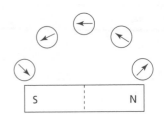

⑤ On the paper, mark a new point where the needle is.
⑥ Repeat and connect the marks until the full field is plotted.

Electromagnetism and Solenoids

- Whenever a current flows in a conducting wire, a magnetic field is produced around the wire.
- The direction of the field lines depends on the direction of the current and can be found with the right hand grip method:
  - Grip the wire in your right hand, with the thumb pointing in the direction of the current.
  - The fingers curled around the wire will point in the direction that the field lines should be drawn
- The strength of the field depends on the size of the current and the distance from the wire.
- This effect can be seen by placing a magnetic compass at different points along the wire and turning the power supply on an off.
- A solenoid is formed when a wire is looped into a cylindrical coil.
- Shaping the wire into a solenoid increases the strength of the magnetic field, creating a strong uniform field inside the solenoid.
- To increase the field strength further, an iron core can be added. This creates an electromagnet.
- The solenoid increases the magnetic field strength because:
  - it concentrates a longer piece of wire into a smaller area
  - the looped shape means that the magnetic field lines around the wire are all in the same direction.
- The magnetic field around a solenoid has a similar shape to that around a bar magnet.
- The north pole of a solenoid can be found with the right hand grip method:
  - Hold the solenoid in your right hand with your fingers following the direction the current flows.
  - Your thumb will point to the north pole of the solenoid.

Right Hand Grip Method

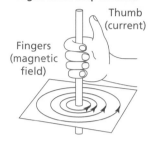

Thumb (current)

Fingers (magnetic field)

A Solenoid

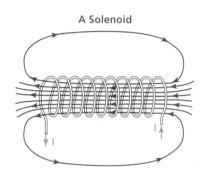

Right Hand Grip Method

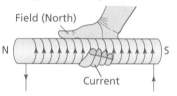

Field (North)

N   S

Current

Electromagnetic Devices

- Many devices use electromagnets.
- You need to be able to interpret diagrams to determine how they work. For example, here is an electric bell:
  ① When the switch is pushed, the electromagnet is magnetised.
  ② The electromagnet attracts the armature.
  ③ The hammer strikes the gong and breaks the circuit.
  ④ The armature springs back, completing the circuit again and remagnetising the electromagnet.
  ⑤ The cycle repeats for as long as the button remains pushed.

Electric Bell

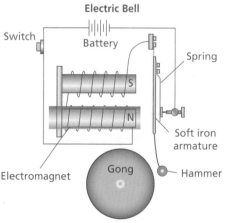

Switch

Battery

Spring

S

N

Soft iron armature

Electromagnet

Gong

Hammer

Quick Test

1. The field lines around a magnet are drawn and the lines are very close together. What does this tell you about the strength of the magnet?
2. Sketch the field lines around a current carrying straight wire.
3. How could the strength of an electromagnet be increased?

Key Words

pole
permanent magnet
induced magnet
flux density
solenoid
uniform
electromagnet

The Motor Effect

You must be able to:

- HT Recall and use Fleming's left hand rule
- HT Calculate the force felt by a current carrying conductor in a magnetic field
- HT Explain how rotation is caused in an electric motor
- HT Explain how loudspeakers and headphones work.

Fleming's Left Hand Rule

- When a current carrying conductor is placed in a magnetic field it experiences a force. This is called the motor effect.
- The motor effect is caused by the field created by the current interacting with the magnetic field.
- The force can be increased by increasing either:
  - the size of the current
  - the length of conductor in the magnetic field
  - the flux density (see page 66).

Creating a Current

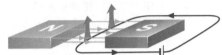

force = magnetic flux density × current × length (of wire within the field)
$F = BIl$

Force (F) is measured in newtons (N).
Magnetic flux density (B) is measured in tesla (T).
Current (I) is measured in amps (A).
Length (l) is measured in metres (m).

A 10cm length of wire with a 4A current flowing passes through a magnetic field. What magnetic flux density is needed to create a 2N force on the wire?

$F = BIl$

$B = \dfrac{F}{Il}$ ← Rearrange the equation to make B the subject.

$B = \dfrac{2}{4 \times 0.1} = \dfrac{2}{0.4} = 5T$

Convert the length from cm to m.

- Reversing the direction of either the current or the magnetic field will cause the direction of the force to reverse.
- The direction of the force on the conductor can be found using Fleming's left hand rule:
  1. Hold the left hand so that the thumb, first finger and second finger are all at right-angles to one another.
  2. Align the first finger, so that it points in the direction of the magnetic field, from north to south.
  3. Rotate the wrist so that the second finger points along the wire in the direction the current is flowing.
  4. The thumb will be pointing in the direction of the force, i.e. the direction the conductor would move.
- The left hand rule can be used to find the direction of either the field, current or movement as long as the other two are known.

Left Hand Rule

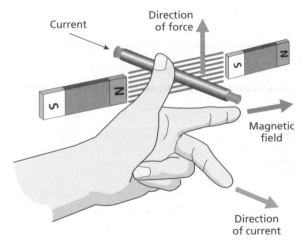

Electric Motors

- A current carrying coil in a magnetic field will rotate.
- This is because the current going up one side of the coil is in the opposite direction to the current coming back down the other side, so one side moves up and the other moves down.
- This is the basis of an electric motor, which is designed so that the coil rotates continuously.

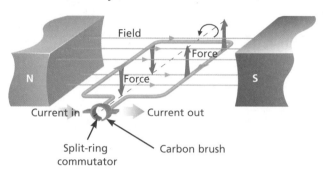

Revise

HT ▶ Key Point

The left hand rule can be remembered by:

- First finger = Field
- Second = Current
- Thumb = Movement

You must remember that for the motor effect, it is the left hand rule.

- The brush contacts at the commutator ensure that the current direction in the coil is always in the same direction.
- This ensures that the motor continues rotating and does not simply stop in the upright position.
- Fleming's left hand rule can be used on one side of the coil to work out which direction the motor will rotate.
- Increasing the current or the magnetic field will make the motor rotate faster.
- Reversing the current or the magnetic field will make the motor rotate in the opposite direction.

Loudspeakers

- Loudspeakers and headphones use the variations in an alternating current to produce sound waves:
  - As the current travels through the coil it experiences a force due to the magnetic field. This makes the speaker cone move.
  - Because the current is alternating, the direction of the force alternates and the speaker cone oscillates.
  - Increasing the electrical power will produce a larger force, so the cone will oscillate with a higher amplitude and produce louder sound waves.
  - The frequency of the sound produced matches the frequency of the alternating current.
  - The size of the speaker affects how quickly or slowly it can oscillate and how big those oscillations will be.
  - A large cone is better for low-pitched sounds and a small cone is better for high-pitched sounds.

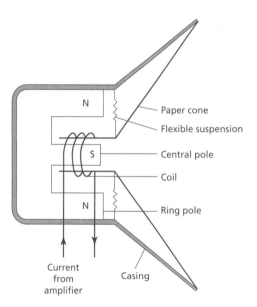

HT ▶ Key Point

The motor effect is used in motors and speakers to convert electrical energy into kinetic and sound energy.

HT ▶ Quick Test

1. How can the speed of a motor be increased?
2. Why will a motor not work without a commutator?
3. How could the pitch of the sound from a speaker be increased?
4. Why must the input signal to a speaker be an alternating current?

HT ▶ Key Words

motor effect
flux density
commutator

Induced Potential and Transformers

You must be able to:

- **HT** Describe the generator effect and apply the principles in a range of contexts
- **HT** Draw and interpret graphs illustrating the potential difference generated by alternators and dynamos
- **HT** Explain how a moving coil microphone and a transformer work
- **HT** Use the transformer equation to make calculations.

The Generator Effect

- When conductors and magnetic fields interact, a potential difference can be induced. This is called the generator effect.
- This potential difference may be induced by:
  - an electrical conductor moving relative to a magnetic field
  - a change in magnetic field around a conductor.
- Increasing the speed of the movement or the size of the magnetic field will increase the size of the induced potential difference.
- Reversing the direction of the movement or the magnetic field will reverse the direction of any induced current.
- An induced current generates a magnetic field that opposes the original change (the movement of the conductor or the change in magnetic field) which produces it.

Alternators and Dynamos

- The generator effect is used in:
  - alternators, to generate an alternating current
  - dynamos, to generate a direct current.
- In an alternator, a rotating magnet is used with a fixed coil of wire.
- As the magnet rotates, the direction of the field through which the coil passes alternates.
- This induces an alternating potential and an alternating current.
- In a dynamo, the coil rotates and the magnet is fixed. As the coil rotates, it generates a potential difference in one direction.
- The use of the split-ring commutator means that once the coil has passed the upright position, the connections are reversed.
- As a result, the direction of the current output is always in the same direction.

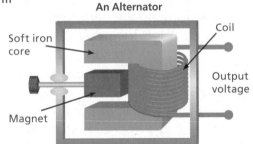

An Alternator

Coil
Soft iron core
Output voltage
Magnet

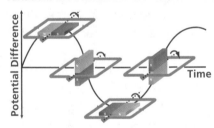

Variations in Potential Difference Over Time

Potential Difference
Time

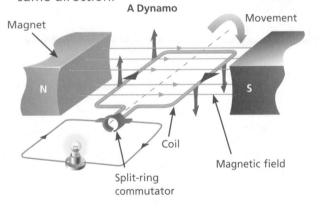

A Dynamo

Magnet
Movement
N
S
Coil
Magnetic field
Split-ring commutator

- In both cases, the induced potential difference is greatest when the magnetic field and the coil are parallel, so the magnetic field is being 'cut' by the rotating coil at the fastest rate.

> **HT Key Point**
>
> Alternators produce alternating current. Dynamos produce direct current.

- During the rotation, at one point the coil and field are at right-angles, so the field lines are not being 'cut' – at this point the induced potential is zero, as can be seen on the graphs.
- In this way, the graphs of potential difference against time can be used to determine the speed of rotation – the time between two zero points is the time for half a rotation.
- For alternators, the speed of rotation is the frequency of the alternating current.

Microphones and Transformers

- A moving coil microphone uses the generator effect to convert sound waves into electrical signals.
  1. The sound waves hit the microphone.
  2. The changes in air pressure related to the sound wave cause the microphone diaphragm to oscillate.
  3. The microphone diaphragm and coil vibrate at the same frequency as the incoming sound wave – the bigger the amplitude of the sound, the bigger the amplitude of vibration.
  4. This induces a potential difference and current in the coil with the same frequency as the incoming sound wave and an amplitude dependent on the amplitude of the incoming wave.
- A basic transformer consists of a primary and secondary coil wrapped around a soft iron core.
- The functioning of a transformer is described below.
  1. An alternating current flows through the primary coil, which is effectively a solenoid (see page 67).
  2. This alternating current induces an alternating magnetic field.
  3. The alternating magnetic field in the iron core induces an alternating potential difference in the secondary coil.
  4. If the secondary coil is part of a complete circuit, an alternating current flows in the secondary coil.
- If transformers were 100% efficient, the power output would be the same as the power input.
- Power = voltage × current, so when the potential difference is increased, the current is reduced and vice versa.

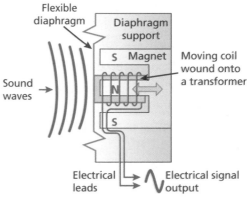

Flexible diaphragm | Diaphragm support
Sound waves
S Magnet — Moving coil wound onto a transformer
N
S
Electrical leads — Electrical signal output

$$\frac{\text{voltage (primary)}}{\text{voltage (secondary)}} = \frac{\text{number of turns (primary)}}{\text{number of turns (secondary)}}$$

$$\frac{V_p}{V_s} = \frac{n_p}{n_s}$$

voltage (secondary) × current (secondary)

= voltage (primary) × current (primary)

$$V_s \times I_s = V_p \times I_p$$

Potential difference (V_p and V_s) is measured in volts (V).

Current (I_p and I_s) is measured in amps (A).
$V_s \times I_s$ = power output (secondary coil)
$V_p \times I_p$ = power input (primary coil)

HT **Key Words**

generator effect
induced
alternator
dynamo
microphone
transformer

HT **Quick Test**

1. How can the size of an induced voltage be increased?
2. What effect would a loud sound have on the potential difference induced by microphone?
3. Describe the difference between a step-up transformer and a step-down transformer in terms of construction and output.

Review Questions

Energy Stores and Transfers

1   On a rollercoaster, the carts are winched to the top of the ride, 16.2m above the start point. Use 10N/kg for gravitational field strength.

   a) The carts and riders on the rollercoaster have a combined mass of 2200kg.

   Calculate the work done by the motor in winching the carts to the top of the ride (assume the winching process is 100% efficient).   [3]

   b) The bottom of the first drop of the rollercoaster is at the same height as the start point.

   Assuming no energy is lost as the carts move down the track, state how much kinetic energy the carts will have when they reach the bottom.   [1]

   c) Use your answer to part **b)** to work out the velocity of the carts at the bottom of the first drop.   [2]

2   For this question you will need to refer to the Physics Equations on page 140.

   A catapult has a spring constant of 100N/m and is stretched 60cm before being fired.

   a) Calculate how much energy has been stored in the catapult.   [3]

   b) Use your answer to part **a)** to work out the velocity of a 57.6g ball bearing fired from the catapult. Assume all of the energy from the catapult goes into the ball bearing.   [3]

   c) When actually measured, it is found that the velocity of the ball bearing is much slower than the calculated value. Suggest why this might be.   [2]

Total Marks _____ / 14

Energy Transfers and Resources

1   Complete **Table 1** to show the main energy transfers that occur in the situations given.

Table 1

Input Energy	Situation	Useful Output Energy
	Walking uphill	
	Sliding down a slide	
	Firing a catapult	
	Fireworks / sparklers	

[4]

2   As household solar panels have become cheaper, they are becoming more commonplace in the UK, although their output is still relatively low.

a)   Manufacturers of solar panels say it doesn't matter that the output is low, because the energy is free. Suggest what they mean by this.   [2]

b)   Other people disagree that the energy is free. Suggest what reasons they would give for their opinion.   [2]

> Total Marks .... / 8

Waves and Wave Properties

1   Write down the wave equation that links frequency, velocity and wavelength.   [1]

2   a)   Waves can be made on a spring by holding one end and moving it up and down, as shown in **Figure 1**.

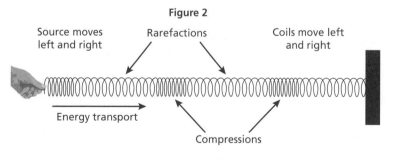

Figure 1

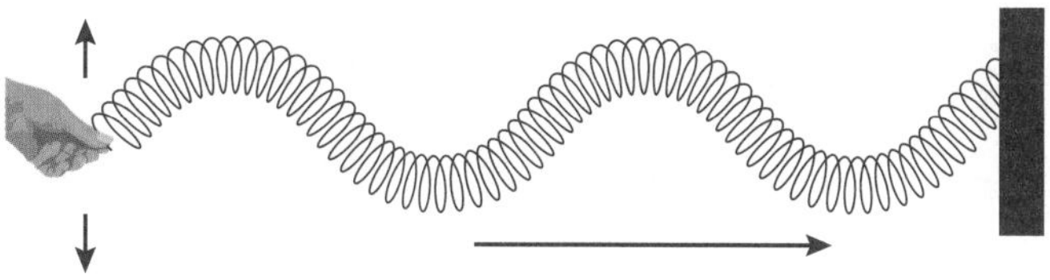

i)   What type of wave is shown in **Figure 1**?   [1]

ii)   How could the person producing the wave increase the frequency of the wave?   [1]

iii)   How could the person producing the wave increase the amplitude of the wave?   [1]

b)   Waves can also be made by moving a spring backwards and forwards, as shown in **Figure 2**.

Figure 2

Source moves left and right   Rarefactions   Coils move left and right

Energy transport

Compressions

i)   What type of wave is shown in this diagram?   [1]

ii)   Give an example of a wave that travels in this way.   [1]

c)   What is transferred by all waves, regardless of the type of wave?   [1]

> Total Marks .... / 7

Reflection, Refraction and Sound

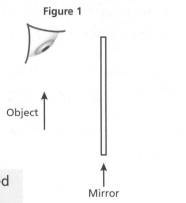

Figure 1

1. Add ray lines to **Figure 1** to show:

   • The formation and position of the object.

   • How the person sees the image. [4]

Object

Mirror

2. HT Use the wave model to explain why light is refracted as it travels from one medium to another. [4]

3. A student carries out an investigation into how the angle of incidence affects refraction by a glass block.

   Give a variable that should be controlled and explain why it is important. [3]

4. HT Explain how a sound wave travelling through a solid can cause the solid to vibrate. [3]

5. HT Give **two** examples of sound waves being converted into vibrations. [2]

Total Marks _____ / 16

HT Waves for Detection and Exploration

1. Describe how a prenatal ultrasound scan is able to detect the foetus in the mother's womb. [3]

2. a) Give **one** similarity between primary seismic waves (P-waves) and secondary waves (S-waves). [1]

   b) Describe **two** differences between primary seismic waves (P-waves) and secondary waves (S-waves). [2]

3. During an earthquake, three different monitoring stations pick up the P-waves and S-waves.

   At station A, the P-waves and S-waves arrive at almost the same time.
   At station B, the S-waves arrive a long time after the P-waves.
   At station C, P-waves are detected but S-waves are not.

   Use this information to describe the positions of stations A, B and C in relation to the epicentre of the earthquake. [4]

4　On **Figure 1**, sketch the path of S-waves through the Earth's surface and label the S-wave shadow zone. [3]

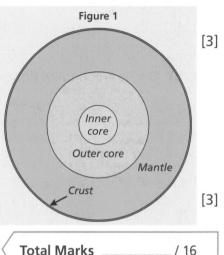

Figure 1

5　A pilot ship uses ultrasound to measure the depth of a river, to determine if the river is deep enough for a larger ship to travel up.

The pilot ship sends an ultrasound pulse downward and detects the reflected pulse 0.04 seconds later.

How deep is the river (speed of sound in water = 1500m/s)? [3]

Total Marks .... / 16

The Electromagnetic Spectrum

1　When electromagnetic waves are absorbed, they transfer energy to the object absorbing them.

Describe how this enables the following processes to occur.
You must name the type of electromagnetic wave that is carrying the energy in each case.

a)　Plants are able to grow. [2]

b)　Bread is toasted when put into a grill. [2]

c)　Television signals are received by an aerial. [2]

2　Complete the sentences using words from the box.

bacteria	high	most	low	least	viruses

Gamma rays have a very .... frequency and are the .... energetic

of all electromagnetic waves. They can be used to destroy .... [3]

3　X-rays are an ionising radiation.
They can be used to treat cancer. However, exposure to X-rays can also cause cancer.

Explain what is meant by 'ionising radiation' and how it can cause cancer. [4]

4　**HT** What type of current is produced when an electromagnetic wave is absorbed by a conductor? [1]

Total Marks .... / 14

Lenses

1 a) On **Figure 1**, draw light rays to show the formation of an image by a concave lens.

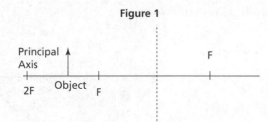

Figure 1

[3]

b) On **Figure 2**, draw light rays to show the formation of images by convex lenses.

Figure 2

Distant object (beyond '2F')

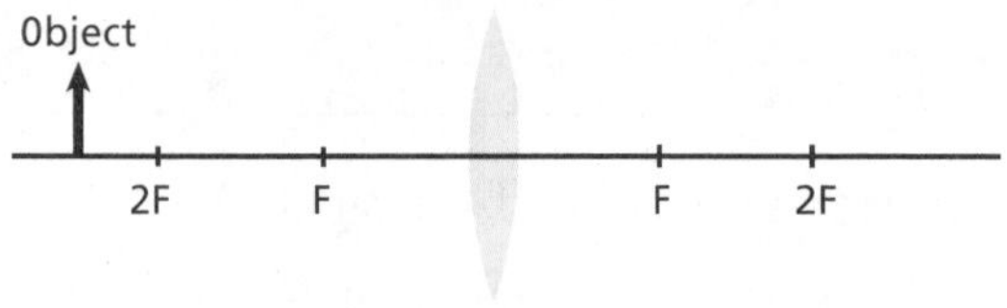

Object between 'F' and '2F'

[6]

2 Explain the difference between a real image and a virtual image. [3]

Total Marks .... / 12

Light and Black Body Radiation

1 Name the **three** things that can occur when light is incident on an object. [3]

2 A child's toy has blue and red stripes.

a) What does this mean about the colours and wavelengths of light that it reflects and absorbs? [3]

b) The toy is viewed through a piece of blue stained glass.

What colour or colours will the blue and red stripes on the toy appear to be? [2]

3 Explain how a black object is different from a white object in terms of the colour and wavelengths of light reflected. [3]

4 How is a perfect black body different from a normal object? [2]

5 _{HT} The graph in **Figure 1** shows how the peak frequency and intensity of radiation emitted from a black body varies with temperature.

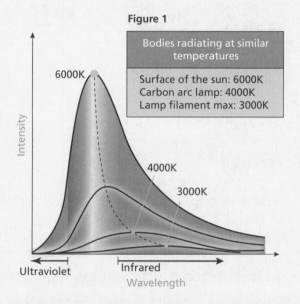

Figure 1

a) Use the graph to explain why a star appears white but a hob on a cooker appears red. [3]

b) Use the graph to explain why objects can still be hot enough to cause burns even if they are not 'glowing' with heat. [3]

c) Some of the hottest stars in the universe are blue super giants.

Use the graph to estimate the temperature of a blue super giant. [1]

Total Marks _____ / 20

An Introduction to Electricity

1. Draw a diagram to show the circuit you would use to measure the voltage across a bulb and the current through the bulb when connected to a battery. [4]

2. Name the component represented by each circuit symbol.

   a) ⊣⊢ [1]

   b) ⎓ [1]

   c) ⊳| LED [1]

   d) ⎓ [1]

3. A bulb is set up in a circuit.
   When there is a potential difference of 12V across the bulb, a current of 0.5A flows.

   a) Calculate the resistance of the bulb. [3]

   b) The bulb is changed for a different type of bulb with half the resistance.

      If the voltage remains unchanged, how will this affect the current that flows? [1]

4. Some students investigate the factors that affect the resistance of a light dependent resistor.
   One student uses a circuit containing a variable resistor connected to a 12V supply and adjusts the resistor to set the output voltages.
   Another student uses a lab pack with outputs 2V, 4V, 6V, 8V, 10V and 12V to set their voltages.

   a) What advantage could using a variable resistor have over using a lab pack with a range of fixed outputs? [2]

   b) During this experiment, the voltage and current across the resistor are measured.

      Why is it important to keep the current low? [2]

   c) **Table 1** shows the results of the experiment.

      Table 1

Light Level	Voltage	Current
Total darkness	10V	0.2A
Low light	6V	0.3A
Maximum brightness	0.4V	0.4A

      Use the results to calculate the maximum change in resistance of a light dependent resistor. [2]

   Total Marks _____ / 18

Circuits and Resistance

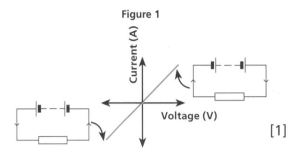

Figure 1

1 **Figure 1** shows the *V–I* graph for an ohmic conductor.

   a) What can be deduced from **Figure 1** about the
      resistance of an ohmic conductor? [1]

   b) Add a second line to **Figure 1** to represent a different
      ohmic conductor with a higher resistance than
      the original. [2]

2 A circuit is constructed with a battery and LED.
  The current through the LED is found to be 1.2A.

  What would the current through the LED be if the battery was reversed? [1]

  Total Marks _____ / 4

Circuits and Power

1 a) Draw a series circuit with a 6V battery and three bulbs. [2]

  b) Each bulb in your series circuit is identical.

     Work out the potential difference across each bulb. [1]

  c) The current through the battery is 2A.

     Work out the current through each bulb. [1]

  d) Use your answers to parts **a)** and **b)** to calculate the resistance of each bulb. [3]

  e) What is the total resistance of the circuit? [1]

2 a) In a series circuit, what effect does adding additional resistors have on the total resistance
     of the circuit? [1]

  b) In a parallel circuit, what effect does adding additional resistors in parallel have on the
     total resistance of the circuit? [1]

  Total Marks **Practise** _____ / 10

Practice Questions

Domestic Uses of Electricity

1. Name the **three** wires in a UK mains cable and, for each one, give its colour. [3]

2. Explain why touching a live wire will cause a large current to flow through the person touching it. [4]

   You must refer to electrical potential and potential difference in your answer.

3. What is the function of a neutral wire in a circuit? [1]

4. In a plug, at what potential is the earth wire? [1]

Total Marks .... / 9

Electrical Energy in Devices

1. Here is some information about a vacuum cleaner for sale in a shop:

   Super Vac: 230V, 1000W Power, extra quiet, light weight, £120

   a) What is power a measure of? [1]

   b) Calculate the current supplied to the vacuum cleaner when it is working at maximum power. [3]

2. Write down the formula that links power, current and resistance. [1]

3. A circuit containing a 5kΩ resistor has a current of 0.1A running through it.

   a) Calculate the power being used by the resistor. [3]

   b) The circuit is switched on for two minutes.

      Calculate the energy transferred by the resistor in this time. [3]

4. **Table 1** gives the power rating and daily usage for a number of different appliances.

Table 1

Appliance	Power Rating	Total Hours Used Per Day
Kettle	2000W	1
Tumble Dryer	1800W	3
Television	500W	6
Oven	5000W	1

a) Which appliance uses the least energy per hour? [1]

b) Work out which appliance uses the most energy during one day. [2]

c) Calculate how many joules of energy are used by the oven per day. [4]

d) The oven has a power supply of 240V.

   Use your answer to part c) to calculate the charge that has been transferred through the oven in a day.

   energy transferred = charge × potential difference [2]

5 Explain why the national grid transfers electricity at a high voltage and low current. [2]

> **Total Marks** _____ / 22

Static Electricity

1 A balloon is rubbed on a jumper and becomes negatively charged.

a) Explain in terms of particle movement how the balloon becomes negatively charged. [3]

b) The balloon is moved close to some small pieces of ripped up paper.

   What would you expect to see happen? [1]

2 A Van de Graaff generator is used in a classroom to demonstrate static electricity.

a) A student stands on an insulated tile with her hand touching the top of the generator. Her hair stands up and sticks out.

   Explain why this occurs. [3]

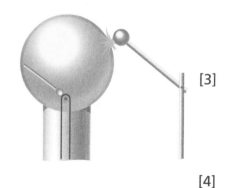

b) An earthed rod is brought close to the top of the generator. A spark jumps across from the generator to the earthed rod.

   Explain why this occurs. [4]

3 On an electric field diagram, what does the direction and density of the arrows on the field lines indicate? [1]

> **Total Marks** _____ / 12

Magnetism and Electromagnetism

1 Draw the magnetic field lines around a bar magnet. [3]

2 Explain what is meant by 'induced magnetism'. [1]

3 Some students are carrying out an experiment with plotting compasses and electric circuits. They notice that the compass needle moves when the circuit is switched on, indicating that the current produces a magnetic field.

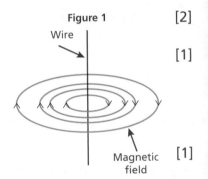

Figure 1

a) Give **two** ways in which the size of this magnetic field could be increased. [2]

b) How could the direction of the magnetic field be reversed? [1]

c) **Figure 1** shows the magnetic field lines around the wire.

Draw an arrow on the wire in **Figure 1** to show the direction of the current through the wire. [1]

Total Marks .... / 8

HT The Motor Effect

1 **Figure 1** shows a conducting wire carrying a current placed in a permanent external magnetic field.

The field from the wire interacts with the field from the magnet causing the wire to move.

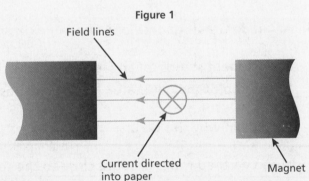

Figure 1

a) What is the name of the rule that can be used to determine the direction of this movement? [1]

b) Use this rule to determine the direction in which the wire in **Figure 1** moves. [1]

c) Give **two** examples of where this motor effect is used. [2]

d) Give **one** way in which the size of the force can be increased. [1]

e) Give **one** way in which the direction of the force can be reversed. [1]

2 The strength of the Earth's magnetic field is around 50 micro teslas (50 × 10⁻⁶T).

Calculate the force experienced on a 100m radio transmitter that has a 12A current flowing through it.
Refer to the Physics Equations on page 140.

[2]

3 An electric motor contains a coil of wire.

Each turn has a 2cm length in the magnetic field and the magnetic field strength is 0.3T.
Refer to the Physics Equations on page 140.

a) Calculate the force on each turn when a current of 1A flows. [2]

b) The coil on the motor contains 200 turns of wire.

Calculate the total force on the coil. [1]

Total Marks _____ / 11

HT Induced Potential and Transformers

1 The generator effect is also used in transformers.
Figure 1 shows a transformer.

a) Label parts **i)**, **ii)** and **iii)** on **Figure 1**. [3]

b) Determine whether the transformer in **Figure 1**
is a step-up or step-down transformer. [1]

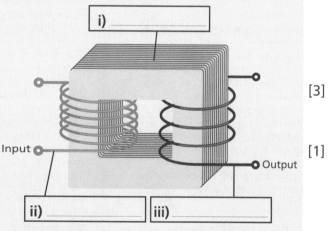

Figure 1

i) _____

Input

ii) _____   iii) _____

Output

c) A step-down transformer is being used in a laptop computer.
It uses the mains supply at 230V to produce an output potential difference of 12V.

If there are 4600 turns on the largest coil, calculate the number of turns on the other coil.
Refer to the Physics Equations on page 140. [3]

Total Marks _____ / 7

Particle Model of Matter

You must be able to:

- Draw simple diagrams to model the different states
- Describe how to investigate the density of objects
- Explain what is meant by 'internal energy' and how this links to temperature and changes of state
- Use the particle model to explain pressure in gases
- Explain how pressure, volume and temperature are interrelated.

States of Matter

- There are three states of matter: solids, liquids and gases.
- Solids and liquids are incompressible (cannot be squashed) because there are no gaps between the particles in them.
- Solids contain particles in a fixed pattern and have a fixed size and shape.
- Liquids have a fixed size but contain particles that are free to move, allowing them to change shape to fit their container.
- Gases have large gaps between the particles, making them compressible and enabling them to change size and shape.
- The density of a material is its mass per unit volume.

Gas

Liquid

Solid

Temperature decreases

 LEARN

$$\text{density} = \frac{\text{mass}}{\text{volume}} \qquad \rho = \frac{m}{V}$$

Density (ρ) is measured in kilograms per metre cubed (kg/m³).

Mass (m) is measured in kilograms (kg).

Volume (V) is measured in metres cubed (m³).

REQUIRED PRACTICAL
Investigate the density of regular and irregular solids and liquids.

Sample Method	**Considerations, Mistakes and Errors**
1. Set the equipment up as shown. 2. Record the height of the water in the measuring cylinder and the mass of the solid / liquid being tested. 3. Add the solid / liquid being tested to the measuring cylinder. 4. Record the new height in the measuring cylinder. 5. Subtracting the original height from the new height gives the volume of the solid / liquid being tested. 6. Now the density can be calculated.	• If a solid that is less dense than water is tested, the volume measurement will be incorrect because the solid will not be fully submerged. • When reading from the measuring cylinder, the reading should be taken from the bottom of the meniscus. • The temperature of the water must be exactly the same throughout all tests, as an increase in temperature could cause the material or water to change volume slightly through expansion.
Variables • The independent variable is the material being tested. • The dependent variables are the volume and mass. • The control variable is the temperature.	**Hazards and risks** • There are very few hazards, unless the materials being tested are hazardous or react with water. • The main hazard could be a slip hazard if water is spilt.

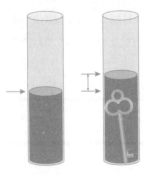

Change of State and Specific Latent Heat

- Internal energy is the total kinetic energy and potential energy of all the particles in a system and is stored by the particles.
- As energy is added to a system, its temperature will increase.
- At the melting or boiling point, the energy being added causes the substance to change state not temperature.
- A change of state is reversible.
- It is a physical change that alters the internal energy but not the temperature or mass. It is not a chemical change.

> **Key Point**
>
> The energy needed to cause 1kg of a substance to change state is called its specific latent heat.

$$\text{energy for a change of state} = \text{mass} \times \text{specific latent heat}$$
$$E = mL$$

- The **latent heat of fusion** is the energy needed for a substance to change from solid to liquid (melt).
- The **latent heat of vaporisation** is the energy needed for a substance to change from liquid to gas (evaporate).
- The horizontal parts of heating and cooling graphs indicate where energy is being used to change state.

Energy (E) is measured in joules (J).
Mass (m) is measured in kilograms (kg).
Specific latent heat (L) is measured in joules per kilogram (J/kg).

Melting and Boiling

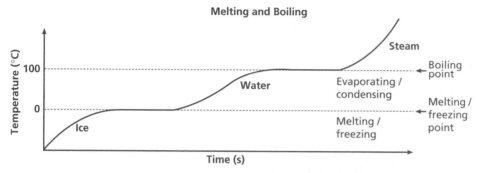

Gas Inside a Piston

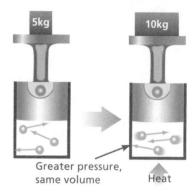

Greater pressure, same volume

Heat

Particle Motion and Pressure in Gases

- The particles of a gas are in constant random motion and its temperature is related to the average kinetic energy of the particles.
- When the particles of a gas collide with the walls of their container they exert a force on the wall, which is felt as pressure.
- If the volume is kept constant, increasing the temperature increases the speed of the particles. This increases the frequency and force with which the particles hit the walls and increases the pressure.
- Changing the volume of a gas can also change the pressure.
- If the temperature is kept constant:
  - increasing the size of the container will reduce the pressure, as the particles will collide with the walls less frequently
  - reducing the volume will increase the pressure, as the particles will collide with the walls more frequently.
- For a fixed mass of gas at constant temperature $p_1V_1 = p_2V_2$

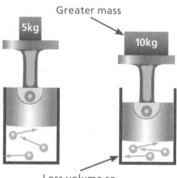

Greater mass

Less volume so higher pressure

$$\text{pressure} \times \text{volume} = \text{constant}$$
$$pV = \text{constant}$$

Pressure (p) is measured in pascals (Pa).
Volume (V) is measured in metres cubed (m³).

HT When work is done on a gas, it increases the internal energy of the gas and can cause an increase in temperature.

HT When a bicycle pump is compressed, the piston hits the gas particles giving them energy so the temperature increases.

HT As they are moving faster, the particles collide with objects more frequently and the pressure will increase.

> **Key Words**
>
> state of matter
> incompressible
> particles
> density
> meniscus
> latent heat of fusion
> latent heat of
>   vaporisation

Quick Test

1. Why are solids and liquids incompressible?
2. How could you find the density of an irregular solid?
3. Why would an increase in temperature affect the pressure of a gas?

Atoms and Isotopes

You must be able to:

- Recall the approximate size of an atom
- Describe the structure of an atom
- Explain what an isotope is
- Describe how isotopes of the same element are different
- Describe the plum pudding model of the atom
- Describe evidence that has led to the nuclear model of the atom.

The Structure of the Atom

- Atoms are very small with a radius of around 1×10^{-10} metres.
- Atoms contain a positively charged nucleus made up of protons and neutrons, which is surrounded by negatively charged electrons.
- Protons have an electrical charge of +1 and electrons have a charge of –1.
- Atoms have equal numbers of electrons and protons, so they have no overall electrical charge.
- The nucleus of the atom contains most of the mass, but its radius is less than $\frac{1}{10\,000}$ of the radius of the atom.
- The electrons are arranged at different distances from the nucleus (in different energy levels).
- The energy level of an electron may change when the atom emits or absorbs electromagnetic radiation:
  - Absorbing electromagnetic radiation moves electrons to a higher energy level, further from the nucleus.
  - Electromagnetic radiation is emitted when an electron drops to a lower energy level.
- An atom that loses one of its outer electrons becomes a positive ion.
- If it gains an extra electron, it becomes a negative ion.

Isotopes

- All atoms of a particular element have the same number of protons.
- The number of protons in an atom of an element is called its atomic number.
- The total number of protons and neutrons in an atom is called its mass number.
- Atoms are represented in the way shown opposite.
- Atoms of the same element can have different numbers of neutrons.
- These atoms are called isotopes.
- For example, carbon has two common isotopes:
  - carbon-12, which contains 6 protons and 6 neutrons
  - carbon-14, which contains 6 protons and 8 neutrons.
- These isotopes are represented with the following symbols:

$$^{14}_{6}\text{C} \qquad ^{12}_{6}\text{C}$$

- Note that in both examples, the atomic number is 6 but the mass number is different.

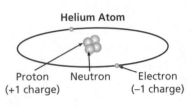

Helium Atom

Proton (+1 charge) Neutron Electron (–1 charge)

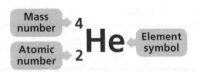

Mass number → 4
Atomic number → 2 **He** ← Element symbol

Key Point

All isotopes of an element contain the same number of protons, otherwise they would be different elements. It is the number of neutrons that is different.

The Plum Pudding Model

- The model of the atom has changed over the years.
- Atoms were once thought to be tiny spheres that could not be divided.
- The discovery of the electron by J. J. Thompson, in 1897, led to the plum pudding model of the atom, which depicts the atom as a ball of positive charge with electrons embedded in it, like plums in a pudding.

Plum Pudding Model

Rutherford and Marsden

- In 1905, Rutherford and Marsden bombarded thin gold foil with alpha particles.
- If the plum pudding model was correct, the heavy, positively charged alpha particles would have passed straight through.
- Most particles did pass through, but not all.
- Some of the alpha particles were deflected, so they must have come close to a concentration of charge, unlike the spread out charge described by the plum pudding model.
- Some alpha particles were reflected back, so:
  - they must have been repelled by the same charge that the alpha particles carried
  - the repelling charge must have been much heavier than the alpha particle, or the alpha particle would have passed through.
- The conclusion was that:
  - the mass of the atom was concentrated in a central nucleus, which was positively charged
  - the electrons surround this nucleus.

Rutherford and Marsden Experiment

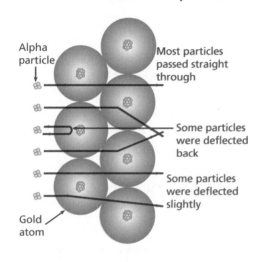

Alpha particle

Most particles passed straight through

Some particles were deflected back

Some particles were deflected slightly

Gold atom

Further Developments

- Niels Bohr adapted this nuclear model by suggesting that electrons orbit the nucleus at specific distances.
- Later experiments led to the idea that the positive charge of the nucleus can be divided into a whole number of smaller particles, each with the same amount of positive charge.
- These particles were given the name 'proton'.
- In 1932, 20 years after the nuclear model was accepted, James Chadwick carried out a number of experiments.
- These provided evidence that within the nucleus there was another particle, which was called the 'neutron', leading to further refinement of the nuclear model for the structure of the atom.
- The structure of the atom is an example of a theory which, through experiment and peer review, has changed and developed over time.

> ### Key Point
>
> New experimental evidence can lead to a scientific model being changed or replaced over time.

> ### Key Words
>
> atom
> nucleus
> proton
> neutron
> electron
> ion
> element
> atomic number
> isotopes

> ### Quick Test
>
> 1. What charge does a proton carry?
> 2. How is an isotope different from an element?
> 3. What does the mass number tell us about an element?
> 4. a) What was the most surprising result from the Rutherford and Marsden experiment?
>   b) What did this result mean?

Nuclear Radiation

You must be able to:

- Explain what is meant by count-rate
- Describe what alpha, beta and gamma radiation are
- Describe the properties of alpha, beta and gamma radiation
- Explain the difference between irradiation and contamination.

Nuclear Decay and Radiation

- Some atomic nuclei are unstable and give out radiation in order to become more stable.
- The type of radiation emitted depends on why the nucleus is unstable and is a random process (it is not possible to predict exactly when an atom will decay).
- The activity of a radioactive source is the rate at which it decays. It is measured in becquerels (Bq).
- One becquerel is equivalent to one decay per second, i.e. 1Bq = 1 decay per second.
- The count-rate is the number of decays recorded each second by a detector (e.g. a Geiger-Muller tube).
- One becquerel is equivalent to one count per second, i.e. 1Bq = 1 count per second.

Alpha, Beta and Gamma Decay

- There are three main types of nuclear radiation: alpha (α), beta (β) and gamma (γ).

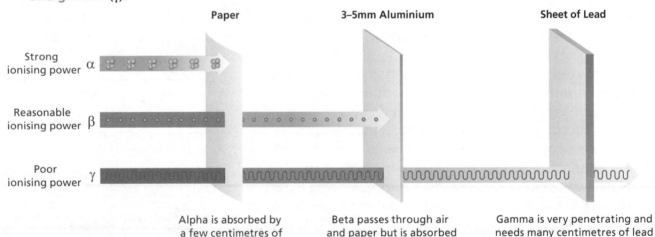

Alpha is absorbed by a few centimetres of air or a thin sheet of paper.

Beta passes through air and paper but is absorbed by a few millimetres of aluminium.

Gamma is very penetrating and needs many centimetres of lead or many metres of concrete to absorb most of it.

- Each type of radiation consists of different particles or electromagnetic radiation.
- All three types of nuclear radiation are ionising – they are capable of knocking electrons from atoms.
- Alpha is the most ionising and gamma is the least ionising.

Radiation Type	Components	Hazards
alpha (α)	• Two neutrons and two protons (the same as a helium nucleus). • Ejected from the nucleus.	• Highly likely to be absorbed and cause damage if passing through living cells.
beta (β)	• A high-speed electron. • Ejected from the nucleus as a neutron turns into a proton.	• Likely to cause damage if absorbed by living cells. • Can penetrate the body to inner organs.
gamma (γ)	• Electromagnetic radiation. • Emitted from the nucleus.	• Likely to pass through living cells without being absorbed and causing ionisation.

- A neutron (n) is the fourth type of nuclear radiation, which can be emitted during radioactive decay.

Radioactive Contamination

- Radioactive contamination is the unwanted presence of materials containing radioactive atoms on other materials.
- The hazard from the contamination is due to the decay of the contaminating atoms.
- The type of radiation emitted affects the hazard.
- Irradiation:
  - is the process of exposing an object to nuclear radiation
  - can be deliberate or accidental
  - does not cause the object to become radioactive.
- When using radioactive sources, it is important to protect against unwanted irradiation by:
  - using sources of the lowest activity possible for the shortest amount of time possible
  - wearing appropriate protective clothing such as a lead apron
  - not handling sources with bare hands.
- You need to be able to compare the hazards associated with contamination and irradiation, e.g.
  - Food contaminated with an alpha source would be more hazardous than food contaminated with a gamma source, because alpha radiation is more strongly ionising.
  - An area contaminated with an alpha source would not be dangerous, unless it was entered, due to the low penetration of alpha radiation. However, if it was contaminated with a source that emitted gamma radiation, this would irradiate people nearby.

Key Point

A contaminated object continues to give out radiation until decontaminated.

An irradiated object does not become radioactive.

Key Words

unstable
activity
radioactive
alpha
beta
gamma
contamination
irradiation

Quick Test

1. Which type of radiation is the most penetrative?
2. Alpha radiation is more ionising than beta radiation. However, it could be argued that beta radiation is more dangerous. Explain why.
3. How is radioactive contamination different from irradiation?

Using Radioactive Sources

You must be able to:

- Work out the half-life of a radioactive isotope from given information
- Describe and evaluate uses of nuclear radiation.

Half-Life

- The random nature of radioactive decay makes it impossible to predict which nucleus will decay next.
- However, with a large enough number of nuclei, it is possible to predict how many will decay in a certain time period.
- The half-life of a radioactive isotope is:
  - the average time it takes for half of the nuclei to decay
  - the time it takes for the count rate, or activity, of a sample containing the isotope to fall to 50% of its original value.
- The graph opposite shows how the count rate of a sample of iodine-128 changes over time.
- The count rate takes 25 minutes to fall from 80 to 40, so its half-life is 25 minutes.

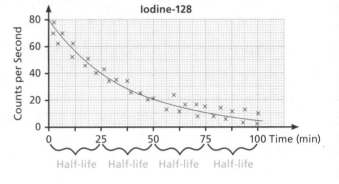

Iodine-128

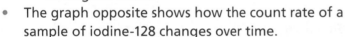
Half-life   Half-life   Half-life   Half-life

HT If the half-life is known, then the activity of a source after a certain amount of time can be calculated.

> The half-life of a radioactive isotope is 2 years and the initial activity is 800Bq. What will be the activity after 6 years?
>
> 2 years = $\frac{1}{2}$ count rate
>
> 4 years = $\frac{1}{4}$ count rate
>
> 6 years = $\frac{1}{8}$ count rate
>
> final count rate = $\frac{1}{8}$ × initial activity
>
> $\qquad = \frac{1}{8} × 800 = 100Bq$

If the half-life is 2 years, then in 6 years' time, 3 half-lives will have passed.

This is the count rate, or activity, after 6 years.

- The half-life of a particular radioactive isotope doesn't change.
- However, the half-life of different isotopes can vary from less than a second to millions of years.
- Isotopes with a short half-life:
  - are very unstable and emit radiation very quickly, so exposure can be very hazardous
  - do not remain radioactive for long.

> **Key Point**
>
> Half-life is a measure of how long it takes for half of the radioactive atoms to decay.

- Isotopes with a long half-life:
  - are more stable and remain radioactive for a very long time
  - emit radiation slowly, so exposure is less hazardous.

Uses of Nuclear Radiation

- Nuclear radiation has many uses, from smoke detectors and industrial uses to numerous medical applications.
- Medical **tracers** are used to look at internal organs:
  - A radioactive isotope is ingested or injected into the body.
  - As it travels around the body, it can be detected on the outside.
- One such use is to monitor kidney function:
  - An isotope that will pass through the kidneys is used.
  - If it builds up in one kidney and not the other, this could indicate that one of the kidneys is not working efficiently.
- Another use could be to look for damage or blockages in the intestines:
  - If there is a blockage, then radioactivity cannot be detected after the blockage.
  - If the intestines are damaged, the radioactive source can be seen to pass out of the intestines into other areas of the body.
- Nuclear radiation can also be used for the treatment of tumours:
  - A tumour in the thyroid gland could be treated with radioactive iodine, which gathers in the gland and destroys nearby cells.
  - A focussed beam of gamma rays can be used to destroy some tumours.
- When choosing which isotope to use for a specific job it is important to consider the half-life, activity and type of radiation.
- A long enough half-life to get results is needed, but it would not be desirable to leave a patient radioactive for a long period of time.
- A gamma source would make the best tracer because:
  - they can penetrate the body and be detected on the outside
  - they are the least ionising.
- Over the years, studies into the effect of radioactivity on humans have enabled scientists to evaluate the risks of using different radiations and compare these to the potential benefits.
- When evaluating these risks, it is important that one scientist's findings are reviewed by others and tested through experiment to prove that they are reproducible.

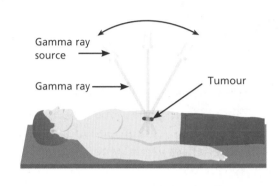

Gamma ray source

Gamma ray

Tumour

> **Key Point**
>
> The use of radiation in medicine is a good example of an application of science where the benefits and hazards need to be evaluated.

> **Quick Test**
>
> 1. Radioactive decay is a random event. Explain what is meant by this.
> 2. The count rate of a radioactive source takes 4 months to fall from 1000Bq to 250Bq. What is the half-life?
> 3. **HT** A radioactive source has a half-life of 3 years. What fraction of the original isotope would remain after 15 years?

> **Key Words**
>
> half-life
> tracer

Fission and Fusion

You must be able to:

- Describe sources of background radiation
- Draw diagrams illustrating a fission chain reaction
- Explain the differences between nuclear fission and nuclear fusion
- Use balanced nuclear equations to illustrate radioactive decay.

Background Radiation

- Background radiation is around us all the time.
- The level of background radiation a person experiences varies depending on their:
  - location, as some areas of the country have a higher amount of natural background radiation
  - occupation, as in some jobs you are more likely to be exposed to radiation.
- Background radiation comes from a range of different sources.

Radon gas
Released at surface of ground
from uranium in rocks and soil

Medical
Mainly X-rays

From food

Nuclear industry

Cosmic rays
From outer space and the Sun

Gamma rays
From rocks and soil
and building materials

13% of radiation is from man-made sources

87% of radiation is from natural sources

Nuclear Fission and Fusion

- Nuclear fission is the splitting of a large unstable nucleus (normally uranium or plutonium).
- Fission will not normally occur by itself – usually, the unstable nucleus must absorb a neutron first.
- During fission, the nucleus:
  - splits into two smaller nuclei of roughly equal size
  - emits two or three neutrons, gamma rays and energy.
- All of the products of fission have kinetic energy.
- The neutrons that are emitted can then go on to start a **chain reaction** by being absorbed by other large, unstable nuclei.

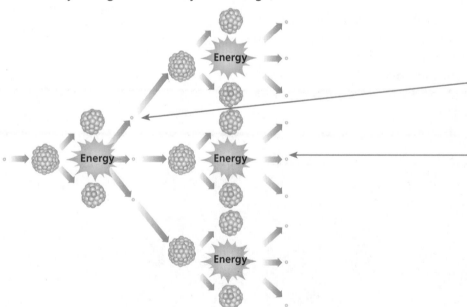

The new neutrons produced can each cause further uranium nuclei to split, so more fission reactions are created. This is a chain reaction, so it carries on and on and on.

The energy is released in the form of heat. Each fission reaction only produces a tiny amount of energy, but there are billions and billions of reactions every second.

- In a nuclear reactor, this chain reaction is controlled to give a constant, steady release of energy.
- The explosion caused by a nuclear weapon is the result of an uncontrolled chain reaction.
- Nuclear fusion can be thought of as the opposite of nuclear fission.
- During fusion:
  - two light nuclei join together to form a heavier nucleus
  - some of the mass is converted into energy and some of this energy may be emitted as radiation.
- Nuclear fusion requires very high temperatures and pressures to:
  - overcome the electrostatic repulsion
  - bring the positive nuclei close enough together for fusion to take place.

> **Key Point**
>
> Nuclear fission is the splitting of large nuclei.
>
> Nuclear fusion is the joining of small nuclei.

Nuclear Equations

- Nuclear equations are used to represent radioactive decay:
  - An alpha particle is represented by the symbol $_2^4He$.
  - A beta particle by the symbol $_{-1}^0e$.
- When an alpha particle is emitted:
  - the mass number of the element is reduced by 4
  - the atomic is number reduced by 2.
- This is because 2 protons and 2 neutrons are emitted from the nucleus, e.g.

$$_{86}^{219}Radon \rightarrow {}_{84}^{215}Polonium + {}_2^4He$$

- With beta decay:
  - the mass number does not change
  - the atomic number is increased by 1.
- This is because a neutron turns into a proton and an electron, and the electron is emitted as the beta particle, e.g.

$$_6^{14}Carbon \rightarrow {}_7^{14}Nitrogen + {}_{-1}^0e$$

- The emission of a gamma ray does not cause a change in the mass or the charge of the nucleus.
- You need to be able to be able to write balanced decay equations for alpha and beta decay:
  - The mass numbers on the right-hand side must add up to the same number as those on the left.
  - The atomic numbers on the right must have the same total as those on the left.
- In the first example above: 219 = 215 + 4 and 86 = 84 + 2
- In the second example above: 14 = 14 + 0 and 6 = 7 − 1

> **Key Point**
>
> How hazardous a type of radiation is depends on its penetrating and ionising power.

> **Quick Test**
>
> 1. Give **one** natural and **one** man-made source of background radiation.
> 2. During a chain reaction, what happens to the emitted neutrons?

> **Key Words**
>
> fission
> chain reaction
> fusion

Stars and the Solar System

You must be able to:

- Describe the formation of the solar system
- Recall and describe the different objects in our solar system
- Explain the balance of forces in a main sequence star
- Describe and explain the lifecycle of a star.

Our Solar System

- The **universe** is made up of billions of galaxies.
- Each **galaxy** contains hundreds of millions of stars.
- Our solar system is a tiny part of the Milky Way Galaxy. It is made up of many objects:
    - one star – the Sun
    - planets and dwarf planets that orbit the Sun
    - asteroids and comets that also orbit the Sun
    - moons that orbit planets – these are referred to as natural satellites.

The Formation of Our Solar System

- The Sun was formed from a nebula (cloud of dust and gas), which was pulled together by gravitational attraction.
- As the dust and gas were drawn together, they collided, increasing the temperature and creating a **protostar**.
- As more and more material was drawn together by gravity, these collisions increased until the temperature and pressure was high enough for hydrogen nuclei to fuse together forming helium and a **main sequence** star.
- The energy released by nuclear fusion keeps the core of the Sun hot.
- Material that was not drawn into the Sun remained in orbit around the new star and formed the planets and other objects in our solar system.
- The Sun is still in the main sequence period of its lifecycle and is stable.
- This stability is the result of the balance between:
    - the fusion energy trying to expand the Sun
    - gravity acting inwards trying to collapse the Sun.

Lifecycle of a Star

- All stars go through a lifecycle that is determined by the size of the star:
    - Small stars, like the Sun, end up as black dwarf stars.
    - Larger stars become neutron stars.
    - The largest stars become black holes.
- All stars begin in the same way as the Sun did – clouds of dust and gas are drawn together by gravity to form a protostar and eventually a main sequence star.

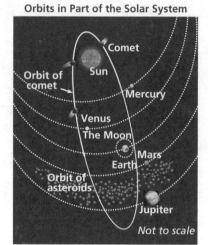

Orbits in Part of the Solar System

Not to scale

Star Formation

- During the main sequence, hydrogen fuses together to create helium and, in large stars, lithium.
- These fusion processes are how new elements are formed.
- As the star ages:
  1. More and more of its mass is converted into energy by nuclear fusion.
  2. As the mass decreases, the outward forces become larger than the force from gravity.
  3. The star then expands and cools becoming a red giant or red super giant.
  4. As the star cools and the hydrogen fuel is used up, the outward forces are reduced.
  5. The star collapses inwards due to gravity and this causes it to increase in temperature once more.
  6. It begins to fuse helium and lithium.
- The largest stars explode as a **supernova**, releasing tremendous amounts of energy and scattering the material of the star into space.

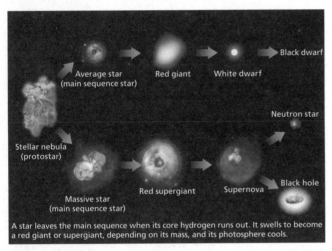

A star leaves the main sequence when its core hydrogen runs out. It swells to become a red giant or supergiant, depending on its mass, and its photosphere cools.

Creating New Elements

- Before stars, the only element in the universe was hydrogen.
- It is fusion processes in stars that have created all of the elements in the periodic table.
- All stars, even the smallest, will fuse hydrogen into helium.
- Bigger stars then fuse helium into lithium and other lightweight elements, up to and including iron.
- During a supernova, the amount of energy released is so great that the temperature and pressure is high enough to force nuclei together to create elements heavier than iron.

> **Quick Test**
>
> 1. What name is given to the first stage in the lifecycle of a star, before fusion begins?
> 2. Sort the following into order of size, smallest first: galaxy, universe, moon, planet, star.
> 3. Which forces are in balance in a main sequence star?
> 4. How does a star form new elements?

> **Key Words**
>
> universe
> galaxy
> nebula
> protostar
> main sequence
> supernova

Orbital Motion and Red-Shift

You must be able to:

- Describe similarities and differences between planets, moons and artificial satellites
- HT Explain why an object stays in orbit
- HT Explain the link between orbital radius and speed
- Explain what is meant by 'red-shift'
- Explain how red-shift provides evidence for the Big Bang theory.

Orbital Motion

- Planets **orbit** the Sun.
- Moons (natural satellites) orbit a planet.
- Artificial satellites are man-made satellites that orbit the Earth, such as those used for satellite TV and GPS systems.
- In all cases, it is the force of gravity that allows planets and satellites to maintain these circular orbits.

HT For an object to travel in a circular path, there must be a force that acts towards the centre of the circle.

HT For planets and satellites, it is the force of gravity that acts towards the centre.

HT This unbalanced force results in an acceleration towards the centre.

HT For an object undergoing circular motion:
- the acceleration does **not** cause the object to change speed
- the acceleration causes the object to change direction
- the velocity changes but the speed remains the same
- the instantaneous velocity is perpendicular to the centripetal force.

HT As the object orbits, the force of gravity pulls the object in a curved path (as the string does in the diagram here). This creates a centripetal force.

HT If an object is in orbit, changing its speed will cause its orbit to change or could cause its orbit to fail.

HT To stay in a stable orbit, at a particular distance from a large **body**, the smaller body must move at a particular speed.

HT If the speed changes, the radius of the orbit must also change.

HT For a stable orbit, the greater the radius the slower the speed.

HT In the diagram here, the geostationary satellite is travelling with a lower velocity than the one in a low polar orbit.

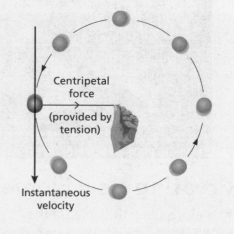

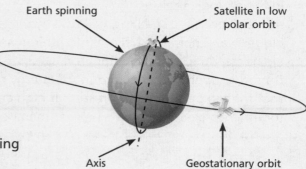

Red-Shift and the Big Bang Theory

- When a source of waves is moving, the waves it produces are:
- squashed together as it moves towards a detector, reducing the detected wavelength

- spread apart as it moves away from a detector, increasing the detected wavelength.
- This effect (called the Doppler effect) is noticeable with sound waves produced by cars – the pitch appears to change if the car is travelling towards or away from an observer.
- The same effect also happens with light.
- If a source of light is moving away from an observer, the wavelength becomes longer and, therefore, redder.
- The faster the source is moving away from an observer:
  - the greater the observed increase in wavelength
  - the greater the red-shift.
- There is an observed increase in the wavelength of light from distant galaxies – the further away the galaxies are, the greater the red-shift.
- This indicates that the more distant the galaxy is, the faster it is moving away from us.
- Red-shift observations indicate that:
  - all the galaxies in the universe are moving away from one another
  - the universe is expanding.
- This provides evidence that the universe is expanding and supports the Big Bang theory.
- The Big Bang theory proposes that the universe began from a very small region that was extremely hot and dense:

- The universe is so vast, and observations so difficult, that there is much that is not known or understood.
- For example, from 1998 onwards, observations of supernovae suggest that:
  - distant galaxies are receding ever faster
  - the rate of expansion of the universe is increasing.
- However, it is not known *how* this increase in expansion is occurring.
- Scientists currently suggest that it is linked to dark matter and dark energy, but little is known about these phenomena and many more observations will be needed before a theory is agreed.

Key Point

An object in orbit has a constant speed but a changing direction – so its velocity is changing.

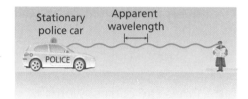
Stationary police car — Apparent wavelength

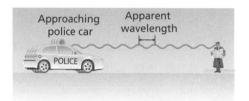
Approaching police car — Apparent wavelength

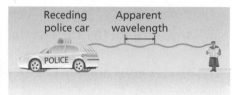
Receding police car — Apparent wavelength

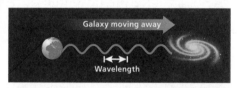
Galaxy moving away — Wavelength

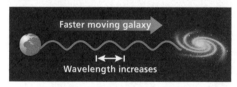
Faster moving galaxy — Wavelength increases

Key Point

Red-shift observations indicate that all galaxies are moving away from one another.

Key Words

orbit
HT circular motion
HT body
red-shift

Quick Test

1. What can be said about the speed of an object in orbit?
2. What force provides the acceleration that causes an object in orbit to follow a circular path?
3. What happens to the observed wavelength of a source that is moving away from an observer?
4. What does red-shift indicate about distant galaxies?

An Introduction to Electricity

1 A bulb has a current of 0.1A flowing through it.

a) Calculate the amount of charge that is transferred if the bulb is turned on for 1 hour. [3]

b) The bulb is changed for a different type of bulb with twice the resistance.

How would the voltage have to change for the same current to flow? [1]

2 Circle the correct words to complete the sentences. [4]

Resistance is a measure of how difficult it is for **voltage / current** to flow. Increasing the resistance means a bigger **voltage / current** is needed if the same **voltage / current** is to flow. A material with a very high resistance is called **a conductor / an insulator**.

3 A light emitting diode has a voltage of 2V across it and transfers 6C of charge per minute.

a) Calculate the current that flows through the LED. [1]

b) Use your answer to part **a)** to calculate the resistance of the LED. [1]

4 Calculate the potential difference needed to push a 2A current through a component with a 12Ω resistance. [2]

> **Total Marks** .... / 12

Circuits and Resistance

1 The current–potential difference graphs for three electrical devices are shown in **Figure 1**.

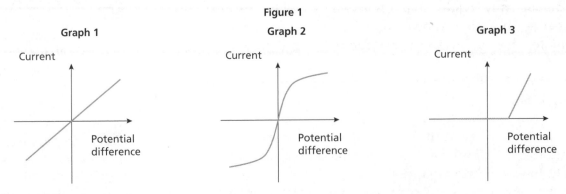

Figure 1

Graph 1   Graph 2   Graph 3

Current / Potential difference (Graph 1)

Current / Potential difference (Graph 2)

Current / Potential difference (Graph 3)

a) Which graph corresponds to each of the components listed below?

i)   A diode. [1]

    ii)   A resistor at constant temperature. [1]

    iii)   A filament lamp. [1]

b) Which graph shows a constant resistance? [1]

c) Which graph shows the resistance increasing as the current increases? [1]

Total Marks _____ / 5

Circuits and Power

1 **a)** Draw a parallel circuit with a 2V battery and two bulbs. [3]

  b) Each bulb in the circuit is identical.

     Work out the potential difference across each bulb. [1]

  c) The current through the battery is 2A.

     Work out the current through each bulb. [1]

  d) Use your answers to parts **a)** and **b)** to calculate the resistance of each bulb. [2]

Total Marks _____ / 7

Domestic Uses of Electricity

1 Give the voltage and frequency of the UK mains electricity supply. [2]

2 **Figure 1** shows a three-pin plug.

Add the following labels to **Figure 1**.

Figure 1

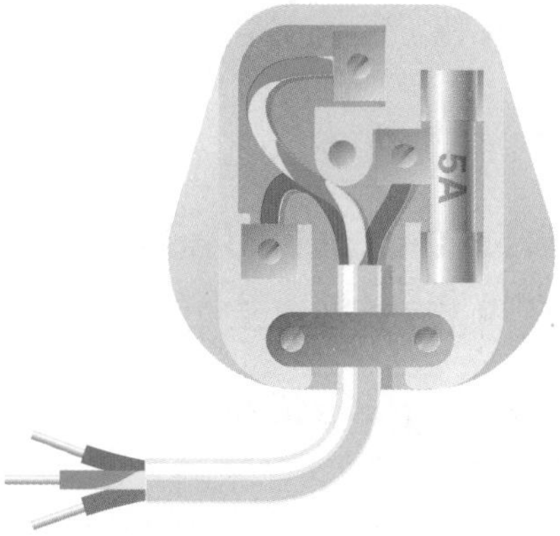

a) Earth Wire [1]

b) Live Wire [1]

c) Neutral Wire [1]

d) Fuse [1]

e) Cable Grip [1]

Total Marks _____ / 7

Electrical Energy in Devices

1 Write down the formula that links power, potential difference and current. [1]

2 An electric heater has a power rating of 2.6kW and is connected to the UK mains supply (230V). The heater is used for 30 minutes.

a) Calculate the current that flows in the heater. [3]

b) Calculate the energy that has been transferred by the heater. [3]

c) Calculate the charge that has been transferred by the heater. [3]

3 **Figure 1** shows the national grid, which is used to distribute electricity from power stations to customers.

Figure 1

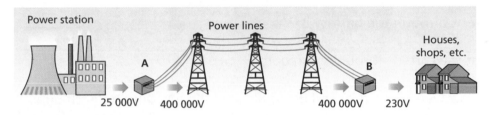

a) What name is given to:

i) Part **A**? [1]

ii) Part **B**? [1]

b) Part **A** increases the voltage.

What effect does this have on the current and why is this important? [2]

c) Describe what part **B** does. [2]

4 Using transformers in the national grid means that energy can be transmitted from the power station to the user with very high efficiency.

Give **three** reasons why this is important. [3]

5 Explain why large steam-powered power stations are more efficient than small steam-powered ones. [3]

Total Marks _____ / 22

Static Electricity

1 Static electricity can be produced when a charge builds up on an insulator or isolated object.

    a) What is meant by an 'isolated' object?   [2]

    b) Why is it not possible to build up charge on a conductor that is not isolated?   [1]

2 When a fuel tanker arrives at a petrol station, the driver always makes sure that the tanker is earthed before pumping fuel.

    Suggest why this is an important safety precaution.   [3]

3 The arrows on an electric field line indicate the direction a positive charge will move in when placed within the field.

Figure 2

On the diagrams in **Figure 2** add the field lines around the charged objects.

The size of the charges is the same.   [4]

+   -   +   +

Total Marks _____ / 10

Magnetism and Electromagnetism

1 A wire is coiled into a solenoid and a current flows through it.

    a) Explain why coiling the wire increases the strength of the magnetic field.   [2]

    b) How could the field strength be increased without adding anything extra to the wire?   [2]

    c) What could be added to the solenoid to further increase the strength of the magnetic field?   [1]

2 On **Figure 1**:

- Mark which end of the solenoid is the north pole.
- Draw the field lines around the electromagnet.   [3]

Figure 1

Review Questions

3 Where around a bar magnet is the field strongest? [1]

4 A student looks at the field lines drawn around two different magnets.
They notice that the field lines around magnet A are much closer together than
those around magnet B.

What does this indicate? [1]

Total Marks _____ / 10

HT The Motor Effect

1 **Figure 1** shows a conducting wire in a magnetic field.

Figure 1

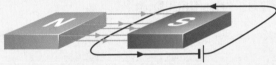

a) Use Fleming's left hand rule to work out the direction the wire will move when
the current flows. Mark this on **Figure 1** with an arrow. [1]

b) The poles of the magnets are reversed.

How will this affect the direction the wire moves? [1]

c) Instead of a battery, the wire is connected to an alternating power supply with
a frequency of 50Hz.

What effect will this have on the movement of the wire? [2]

2 The force experienced by a current carrying wire in a magnetic field depends on
the magnetic flux density.

a) What is meant by flux density? [2]

b) What unit is magnetic flux density measured in? [1]

3 A 4cm length of wire passes through a magnetic field.
It experiences a force of 0.2N when a current of 4A flows through it.

Calculate the strength of the magnetic field.
Refer to the Physics Equations on page 140. [3]

Total Marks _____ / 10

HT Induced Potential and Transformers

1 A dynamo is made from a coil rotating in a magnetic field.

   a) Sketch a graph that shows how the output potential difference varies with time.   [2]

   b) Explain how the shape of your graph depends on the location of the coil as it rotates.   [3]

2 **Figure 1** shows a moving coil microphone.

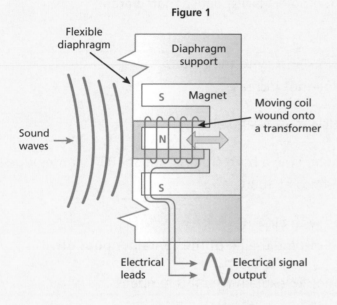

Figure 1

   a) Explain how incoming sound waves are converted into an electrical signal.   [3]

   b) How would increasing the amplitude
of the sound wave affect the output signal?   [1]

3 Explain why a transformer will work with an input alternating current but will not work
with an input direct current.   [4]

> **Total Marks** .... / 13

Particle Model of Matter

1. Solids and liquids are described as incompressible.

   What does this mean? [1]

2. An experiment is carried out to measure the density of an irregular-shaped object.
   The object is submerged in water to calculate its volume.

   Explain why this method will not work if the object is less dense than water. [2]

3. A pan of boiling water is left on the hob.
   The hob remains on and continually transfers energy to the water.
   However, the temperature of the water does not increase.

   Explain why this happens in terms of particle motion and energy. [2]

4. Explain, using the idea of specific latent heat, why a burn from steam at 100°C causes worse injuries than a burn from hot water that is also at 100°C. [2]

5. A sealed container of fixed volume is filled with air.
   The container is heated and after a period of time the lid of the container pops off.

   Use the ideas of pressure and particle motion to explain why this happens. [3]

6. A piston contains a fixed mass of gas.
   Slowly placing masses on top of the piston causes the gas to be compressed without increasing the temperature.
   Initially, the gas is at a pressure of 100 000Pa and a volume 50cm³.
   After the masses have been added, the volume is 40cm³.

   Calculate the pressure of the gas after the masses have been added. [4]

7. **Table 1** contains information about the mass and volume of different objects.
   The density of water is 1g/cm³.

Table 1

	Mass	Volume	Will it Float? Y/N
Object A	60g	40cm³	
Object B	60g	80cm³	
Object C	30g	80cm³	

Calculate the density of each object and determine whether it will float or sink in water. [3]

Total Marks .... / 17

Atoms and Isotopes

1 **Figure 1** shows the structure of a helium atom.

   a) Add labels to **Figure 1** to identify the protons, neutrons and electrons. [3]

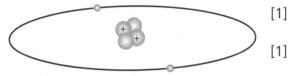

Figure 1

   b) What charge does a proton carry? [1]

   c) Write down the typical radius of an atom. [1]

   d) What would have to happen for the atom to become a negative ion? [1]

   e) Use **Figure 1** to determine the mass number and the atomic number of helium. [2]

2 Carbon-14 is an isotope of carbon with mass number of 14. It contains 8 neutrons.

   Work out the atomic number of carbon. [1]

3 Chlorine has two common isotopes: chlorine-35 and chlorine-37.
   These are represented by the following symbols:

   $^{35}_{17}\text{Cl}$ $^{37}_{17}\text{Cl}$

   a) How many protons does a chlorine atom contain? [1]

   b) How many electrons does a neutral atom of chlorine contain? [1]

   c) Compare the two isotopes of chlorine in terms of the numbers of neutrons in the nucleus. [2]

4 Complete the following sentences using words from the box.

   Each word can be used once, more than once or not at all.

protons	electrons	lower	neutrons	nucleus	energy	radiation	higher

   _____ orbit the _____ of an atom in different energy levels. Absorbing

   electromagnetic _____ can cause _____ to move to a _____ energy level.

   When an electron moves to a _____ energy level, it emits electromagnetic radiation. [6]

5 Describe the plum pudding model of the atom.
   Draw a diagram to support your description. [3]

   Total Marks _____ / 22

Nuclear Radiation

1) Draw **one** line from each type of radiation to the correct description.

Alpha

Weakly ionising – likely to pass straight through living cells without being absorbed.

Beta

Moderately ionising – can pass through the skin and damage organs inside the body.

Gamma

Highest likelihood of being absorbed and causing damage when passing through living cells.

[2]

2) In the 1980s, an explosion at the Chernobyl nuclear power plant released a large amount of radiation into the atmosphere.
Dust containing radioactive particles travelled into the upper atmosphere and fell with the rain onto fields in the UK.

a) Is this type of exposure referred to as contamination or irradiation? [1]

b) After the disaster, sheep farmed in certain areas of the UK were deemed not fit for human consumption.

Suggest why. [2]

3) **Figure 1** shows three different materials.

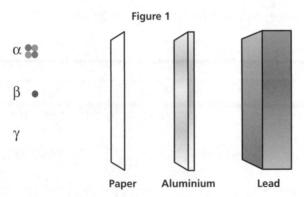

Figure 1

α
β
γ

Paper   Aluminium   Lead

On **Figure 1**, draw a line from each type of radiation to show which of the materials it can penetrate. [3]

Total Marks _____ / 8

Using Radioactive Sources

1 A small sample of a radioactive material is collected and tested in a laboratory.
When first tested the count-rate of the sample is 1000 counts per second.
After 4 minutes, the count-rate falls to 250 counts per second.

  a) Work out the half-life of the sample tested.   [2]

  b) After a few hours, the count-rate remains constant at 8 counts per minute.

   What is causing the count-rate to remain constant?   [1]

  c) A second source was found to have an activity of 150Bq.
   After several months, this activity was still measured to be 150Bq.

   What conclusion can be drawn about the half-life of this source?   [1]

2 There are a number of different uses for radioactive materials and nuclear radiation.

  a) Give **one** medical use.   **b)** Give **one** non-medical use.   [2]

> Total Marks _____ / 6

Fission and Fusion

1 Nuclear fission is a nuclear process that can release a large amount of energy.
For fission to occur, the nucleus must first absorb a neutron.

  a) Describe what happens when fission takes place.   [2]

  b) What type of energy do all products of fission have?   [1]

  c) If the conditions are correct a chain reaction will occur.

   Describe what needs to happen for a chain reaction to occur.   [2]

  d) A chain reaction occurs in both a nuclear reactor and in a nuclear explosion.

   How are the chain reactions in these two situations different?   [2]

2 Nuclear fusion is another method by which energy can be released through nuclear processes.

  Describe the differences between nuclear fusion and nuclear fission.   [4]

> Total Marks _____ / 11

Stars and the Solar System

1 What **two** effects are balanced in a main sequence star? [2]

2 When will a star exit the main sequence phase of its life? [1]

3 What affects the size of the elements that can be produced during a star's main sequence phase? [2]

4 What is the heaviest element that a star can produce without exploding as a supernova? [1]

5 Sort the following sentences into the correct order to explain the processes that take place towards the end of a star's life.

As a star runs out of fuel…

A The mass decreases and the outward forces become larger than the force of gravity.

B The star collapses inwards and the density and pressure increase once more.

C The star expands to become a red giant and starts to cool.

D As the star cools, the forces causing the star to expand begin to fall.

E The increase in density and pressure allows the star to begin fusing heavier elements. [4]

6 This question is about nuclear fusion in stars.
Table 1 gives the atomic masses of some elements.

Table 1

Element	Hydrogen	Helium	Carbon	Nitrogen	Oxygen	Iron	Copper	Zinc
Atomic Mass	1	4	12	14	16	56	64	65

a) Which **two** elements in the table are involved in the fusion process in main sequence stars? [1]

b) Which **two** elements are only produced during a supernova? [1]

c) Heavier elements like oxygen and iron can be formed in the core of a blue super giant. However, they cannot be produced in the core of a main sequence star such as the Sun.

Why is this? [2]

Total Marks _____ / 14

Orbital Motion and Red-Shift

1. Describe the difference between an artificial satellite and a natural satellite and give an example of each. [2]

2. What is the force that keeps satellites in orbit? [1]

3. HT Write down whether each of the statements about circular motion is **true** or **false**.

   a) The object is accelerating towards the centre of the circle. [1]

   b) There is a force acting towards the centre of the circle. [1]

   c) If the speed of the object changes, so will its orbit. [1]

   d) As the radius of an orbit increases, so does the speed of the orbiting object. [1]

   e) The object travels with a constant velocity. [1]

4. A spectator at a Formula One motor race notices that the pitch of the engines appears to change from high to low as the cars go past.

   a) What is the name of this effect? [1]

   b) Explain in detail why the sound of a car moving quickly away will sound lower-pitched. [4]

   c) The same effect is also seen with light.
   The light from a distant galaxy moving away from us appears to have a longer wavelength.

   What is the name given to this effect? [1]

   d) The Andromeda galaxy is moving towards us.

   How will light from the Andromeda galaxy differ from most other galaxies? [1]

   e) During the 1920s, Edwin Hubble discovered that almost all galaxies are red-shifted.

   What does this red-shift show about the movement of the galaxies? [2]

   f) How do red-shift observations show that the universe is expanding? [3]

   g) What theory does this evidence support? [1]

Total Marks .... / 21

Review Questions

Particle Model of Matter

1 All matter is made of particles.

Sketch the particle arrangement for the different states of matter: solid, liquid and gas.
You do not need to draw more than nine particles in each diagram.   [3]

2 **a)** Explain what happens as a solid is heated to become a liquid in terms of particle
movement and energy.   [3]

b) Explain what happens as a liquid is heated to become a gas in terms of particle
movement and energy.   [3]

3 Describe a method that can be used to find the density of an irregular-shaped object.   [4]

4 Energy is provided to a sealed system containing ice.
Figure 1 shows how the temperature of the system changes with time.

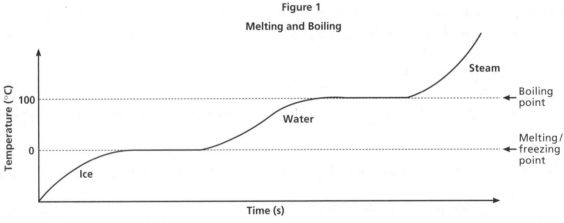

Figure 1
Melting and Boiling

a) Explain the shape of the graph.   [6]

b) The water becomes a gas and the temperature of the system continues to rise.

What effect will this have on the pressure of the gas?   [1]

c) The specific latent heat of fusion of water is 334 000J/kg.

If there was 500g of ice in the system at the start, calculate how much energy is needed to
completely melt the ice once it has reached 0°C.   [2]

Total Marks _____ / 22

Atoms and Isotopes

1. What did J. J. Thompson discover in 1897 that led to the development of the plum pudding model of the atom? [1]

2. In 1905, Rutherford and Marsden carried out an experiment to test the plum pudding model.

   During the experiment a thin gold foil was bombarded with alpha particles and the path of the alpha particles recorded.

   Figure 1

   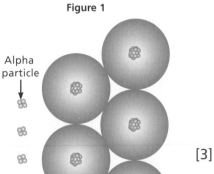

   a) On **Figure 1**, draw three possible paths taken by the alpha particles as they strike the gold foil. [3]

   b) What conclusion did Rutherford make from the path of the alpha particles? [2]

   c) What adaptation did Niels Bohr make to the nuclear model proposed by Rutherford. [2]

3. Uranium is a very heavy atom.
   It has a number of different isotopes:

   - uranium-233 $^{233}_{92}U$
   - uranium-235 $^{235}_{92}U$
   - uranium-238 $^{238}_{92}U$

   a) How many protons does an atom of uranium contain? [1]

   b) Which of the isotopes of uranium is the lightest? [1]

   c) Compare the nuclei of uranium-235 and uranium-238. [2]

   d) When uranium forms ions, it can form U^{3+} and U^{4+} ions.

   How many electrons does a U^{3+} ion have? [2]

   Total Marks _____ / 14

Review Questions

Nuclear Radiation

1. Draw **one** line from each type of radiation to the correct description.

Alpha		An electron emitted from the nucleus
Beta		High-frequency electromagnetic radiation
Gamma		A helium nucleus (2 protons and 2 neutrons)

[2]

2. Explain why it is not possible to predict exactly when a particular radioactive nucleus will decay. [2]

Total Marks _____ / 4

Using Radioactive Sources

1. A doctor is choosing a radioactive isotope to use as a tracer.
   The tracer will be added to a patient's drink so that the doctor can monitor the movement of liquid through the digestive system.
   The doctor intends to use a detector on the outside of the patient's body to see if the radioactive material isotope is building up anywhere.

   Table 1 shows the different isotopes the doctor can choose from.

   Table 1

	Isotope A	Isotope B	Isotope C	Isotope D
Half-Life	4 hours	2 days	3 hours	4 minutes
Type of Radiation	Alpha	Gamma	Gamma	Gamma

   Which isotope should the doctor use?
   You must justify your choice. [3]

2. **HT** Carbon dating is used to determine the age of dead organic materials.
   The process uses radioactive carbon-14, which has a half-life of around 6000 years.

   a) Scientists are studying a mammoth found frozen in ice.
      They find that a sample of carbon from the remains has a count-rate of 200 counts per minute.

The scientists work out that when the mammoth died, the count-rate would have been 800 counts per minute.

Use this information to calculate how long ago the mammoth died. [3]

b) A second set of remains is found that are 18 000 years older than the first set of remains.

If the same size sample is tested, calculate the count-rate of the second sample. [3]

c) Carbon dating can only be used to calculate the age of carbon samples up to 60 000 years old.

Explain why. [2]

Total Marks .... / 11

Fission and Fusion

1. The graph in **Figure 1** shows the decay chain from uranium to thorium.

a) The x-axis states 'Number of Protons'.
It could also have been labelled 'Atomic Number'.

How else could the y-axis have been labelled? [1]

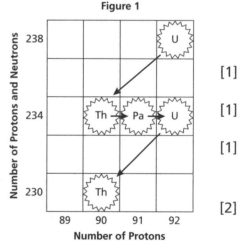

Figure 1

b) What type of decay is indicated by the diagonal arrows? [1]

c) What type of decay is indicated by the horizontal arrows? [1]

d) Complete the decay equation for the decay of uranium-238 into thorium-234. [2]

$$^{238}_{92}U \rightarrow\, ^{....}_{90}U +\, ^{4}_{....}He$$

e) Complete the decay equation for the decay of thorium into palladium. [2]

$$^{234}_{90}Th \rightarrow\, ^{....}_{91}Pa +\, ^{0}_{....}e$$

Total Marks .... / 7

Review Questions

Stars and the Solar System

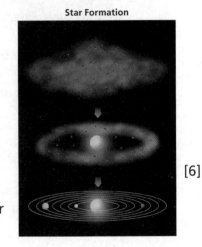

Figure 1

Star Formation

1 **Figure 1** shows the processes that formed our solar system.

Add labels and notes to **Figure 1** to explain what is happening at each of the three stages.

You should mention the forces involved and temperature in your descriptions.

[6]

2 Draw lines to connect each type of star to the two stages that occur towards the end of its lifecycle.

| Stars the size of the Sun | Expands to become a red giant or red super giant. | Explodes in a supernova then contracts to become a neutron star. |

| Larger stars | Expands to become a red giant. | Explodes in a supernova then contracts to become a black hole. |

| Largest stars | Expands to become a red super giant. | Fusion stops and the star becomes a white dwarf and eventually black dwarf. |

[6]

3 Define each of the key terms below:

a) Nebula [1]

b) Protostar [1]

c) Supernova [1]

4 Astronomers may refer to the forces on a star being in equilibrium.

What to do they mean by this and what forces are they referring to? [3]

5 All stars are capable of creating the same element during fusion.

a) What element is this? [1]

b) What element do stars fuse to create this new element? [1]

6 Name the heaviest element that can be fused in the largest stars. [1]

Total Marks .... / 21

Orbital Motion and Red-Shift

1 HT **Figure 1** shows an object undergoing circular motion.

Add **two** straight line arrows to **Figure 1** to show:
- The direction of the force caused by tension on the string.
- The direction of the instantaneous velocity of the orbiting object.

[2]

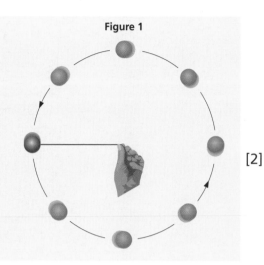

Figure 1

2 Explain what is meant by red-shift. [2]

3 In the 1920s, Edwin Hubble studied the red-shift from a large number of distant galaxies.

He used the red-shift to calculate the speed with which the galaxies were moving away from us (recessional velocity).

The graph in **Figure 2** shows some of his findings.

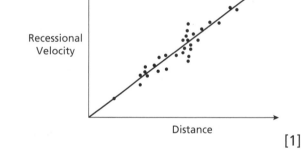

Figure 2

Recessional Velocity

Distance

a) What conclusion can be drawn from the graph? [1]

b) Only a few of the points on the graph fall exactly on the line of best fit.

Why might it have been difficult to get accurate measurements? [2]

c) Describe the Big Bang Theory. [2]

d) Not all people agree with the Big Bang Theory.

Suggest **three** possible reasons. [3]

4 Observations of supernova made in the last twenty years suggest that galaxies are receding even faster.

What does this suggest about the rate of expansion of the universe? [1]

Total Marks .... / 13

1   The graph in **Figure 1** shows the journey of a man walking to the shops.

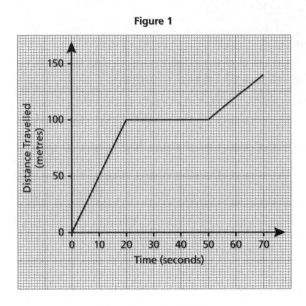

Figure 1

a) Calculate the man's average speed for the whole journey.   [3]

b) Estimate how many times faster the man is travelling in the first part of the graph
   (between 0 to 20 seconds) than in the last part of the graph (between 50 to 70 seconds).   [1]

c) Describe the motion of the man between 20 to 50 seconds.   [1]

2   A skydiver steps out of an aeroplane.
    After 10 seconds, she is falling at a steady speed of 50m/s.

a) What name is given to this steady speed?   [1]

The skydiver opens her parachute.

After another 5 seconds, she is once again falling at a steady speed. This speed is now
only 10m/s.

b) Calculate the skydiver's average acceleration during the time from when she opens her
   parachute until she reaches 10m/s.   [3]

c) Explain as fully as you can:

   i) Why the skydiver eventually reaches a steady speed (with or without her parachute).   [3]

   ii) Why the skydiver's steady speed is lower when her parachute is open.   [2]

d) The skydiver and her equipment have a total mass of 75kg.

   Calculate the weight of the skydiver.   [2]

3 **Figure 2** shows how a simple lever can be used to collect water from a river. A person using the lever applies a force at the end of the lever.

Figure 2

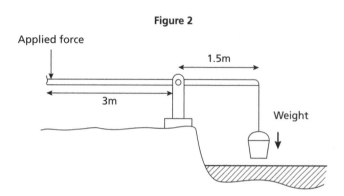

a) The bucket and water have a weight of 200N.

   Calculate the turning force (moment) of the bucket of water. [2]

b) Calculate the downward force a person using the lever would have to apply to lift the bucket. [2]

4 The graph in **Figure 3** shows changes in the velocity of a racing car.

Figure 3

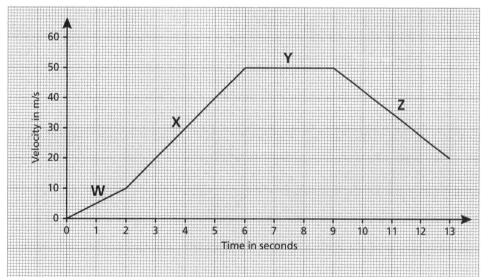

a) Describe the motion of the car in the section marked **Z**. [1]

b) Calculate the acceleration of the racing car during the period labelled **X**. [4]

c) Calculate the distance travelled by the car during the period marked **Y**. [3]

5 The Moon orbits the Earth in a circular path.   [3]

Complete the sentences below using words from the box.

| resistance | speed | velocity | direction |

The Moon's .... is constant. However, its .... changes, because its

.... changes.

6 HT **Figure 4** shows the orbits for two types of satellite: a polar orbit and a geostationary orbit.

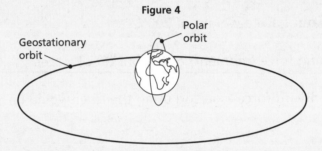

Figure 4

A satellite in stable Earth orbit moves at a constant speed in a circular orbit, because there is a single force acting on it.

a) What is the direction of this force?   [1]

b) What is the cause of this force?   [1]

c) What is the effect of this force on the velocity of the satellite?   [2]

d) In which of the orbits shown in **Figure 4** would this force be bigger?
   You must give a reason for your answer.   [2]

e) Explain why the kinetic energy of a satellite in orbit remains constant.   [2]

7 When a car driver has to react and apply the brakes quickly, the car travels some distance before stopping.
Part of this distance is called the 'thinking distance'.
This is the distance travelled by a car while the driver reacts to a dangerous situation.

Table 1 shows the thinking distance (m) for various speeds (km/h).

Thinking Distance (m)	0	3	6	9	12	15
Speed (km/h)	0	16	32	48	64	80

a) Describe how thinking distance changes with speed.   [2]

b) A driver drank two pints of lager.

Some time later, the thinking time of the driver was measured as 1.0s.

    i) Calculate the thinking distance for this driver if the driver was travelling at 9m/s. [1]

    ii) A speed of 9m/s is the same as 32km/h.

    Use the table to find the thinking distance for a driver who has not had a drink travelling at 32km/h. [1]

    iii) What has been the effect of the drink on the thinking distance of the driver? [1]

8 To get a bobsleigh moving quickly, the crew need to do work to overcome friction. They push the sleigh hard for a few metres and then jump in.

a) Write down the formula used to calculate work done. [1]

b) If the crew exert a force of 400N over 5m, calculate how much work have they done on the sled. [2]

c) As the crew do work on the sled they transfer energy to it.

What type of energy does the sled gain? [1]

d) The energy the crew use to get the sled moving comes from their food.

What type of energy store is this? [1]

9 The Sankey diagrams in **Figure 5** show what happens to each 100 joules of energy stored in coal when it is burned on an open fire and in a stove.

Figure 5

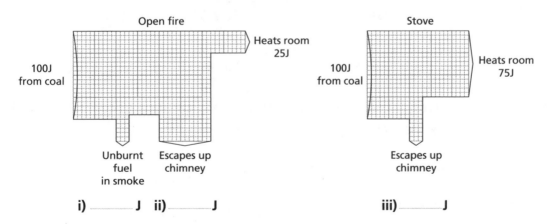

a) Add the **three** missing numbers to the diagrams in **Figure 5**. [3]

b) Work out the efficiency of the open fire. [1]

c) Work out the efficiency of the stove. [1]

10 A cyclist accelerates from a standstill at a set of traffic lights.
The driving force of the back tyre on the ground is 250N.

  a) How much work is done by this force when the cyclist travels five metres? [2]

  b) Describe what happens to the energy transferred by this force. [1]

11 A man's car will not start, so two friends help by pushing it. The car has a mass of 800kg.
By pushing as hard as they can for 12 seconds, the two friends make the car reach a speed of three metres per second.

  a) Calculate the acceleration given to the car. [2]

  b) Whilst pushing the car, the two friends together do a total of 2400 joules of work.

  Calculate their total power. [2]

  c) Another motorist has the same problem.
  The two friends push his car along the same stretch of road with the same force as before.
  It takes them 18 seconds to get the second car up to a speed of three metres per second.

  Calculate the mass of the second car. [4]

12 **Table 2** shows the main sources of energy used for electricity generation in Britain in 2015.

Table 2

Coal	28.2%
Oil and other	2.6%
Gas	30.2%
Nuclear	22.2%
Renewables	16.7%

  a) How does the amount of electricity obtained from nuclear sources compare with the amount obtained from renewables? [1]

  b) Hydroelectricity is a renewable source.

  Name **two** other renewable energy sources. [2]

  c) Give **one** advantage and **one** disadvantage of nuclear power. [2]

d) A hydroelectric dam stores water at a high level.
When the water is released, it flows down pipes to the power station where it turns turbines.

Complete the diagram to show the **useful** energy transfers that occur as the water flows down the pipes **to** the power station.

_____ energy ⟶ _____ energy   [2]

e) The electricity generated a power station is transmitted over long distances.
Before this happens, its voltage is increased by using a step-up transformer.

Describe **one** advantage and **one** disadvantage of transmitting electricity at high voltage.   [2]

13 A forklift truck was used to stack boxes onto a trailer.
It lifted a box weighing 1900N through 4.5m in a time of 9 seconds.

a) Calculate the work done on the box.   [2]

b) Calculate the useful power output of the motor.   [2]

c) The engine and motor combined are 50% efficient.

Calculate the energy used by the forklift truck in lifting the box.   [2]

14 Circuit breakers help to make the electricity supply in homes safer.
A circuit breaker is an automatic safety switch – it cuts off the current if it gets too big.

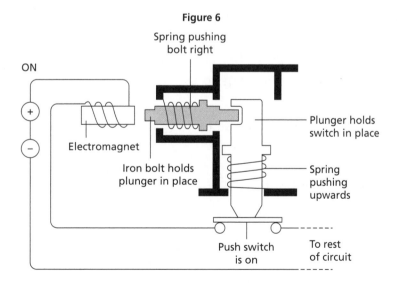

Figure 6

Describe in detail how this circuit breaker show in **Figure 6** works and what happens when the current becomes too big.   [5]

15 An inventor develops a solar-powered bike.
A battery is connected to solar cells, which charge it up.
There is a switch on the handlebars.
When the switch is closed, the battery drives a motor attached to the front wheel.

a) Complete the sentences using words from the box.
Words may be used once, more than once, or not at all.

chemical	electrical	heat (thermal)	kinetic	light
gravitational potential		sound	elastic potential	

i) The solar cells transfer _____ energy to _____ energy. [2]

ii) When the battery is being charged up _____ energy is transferred to

_____ energy. [2]

iii) The motor is designed to transfer _____ energy to _____ energy. [2]

b) The cyclist stops pedalling for 10 seconds.
During this time the motor transfers 1500 joules of energy.

i) Calculate the power of the motor. [2]

ii) Name **one** form of wasted energy that is produced when the motor is running. [1]

16 **Figure 7** shows part of the National Grid.
At **X**, the transformer increases the voltage to a
very high value.
At **Y**, the voltage is reduced to 230V for use by
consumers.

Figure 7

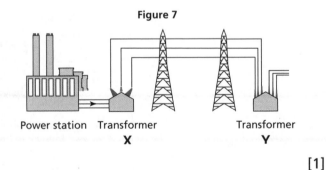

Power station   Transformer   Transformer
X   **Y**

a) What happens to the current as the voltage
is increased at **X**? [1]

b) Why is electrical energy transmitted at very high voltages? [1]

c) What name is given to the type of transformer at **Y**? [1]

17 HT **Figure 8** shows a transformer.

Figure 8

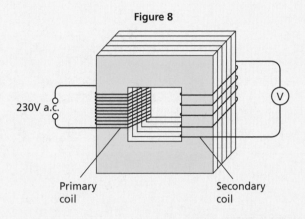

230V a.c.

Primary coil

Secondary coil

a) Name the material used to make the core of the transformer. [1]

b) The primary coil has 46 000 turns and the secondary coil 4000 turns.

Calculate the output voltage if the input voltage is 230V a.c. [3]

18 The information plate on a hairdryer is shown in **Figure 9**.

Figure 9

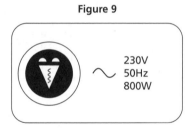

230V
50Hz
800W

a) What is the power rating of the hairdryer? [1]

b) Calculate the current in amps when the hairdryer is being used. [2]

c) Which fuse, 3A, 5A, 13A or 30A, should you use with this hairdryer?
Tick **one** box.

3A ☐

5A ☐

13A ☐

30A ☐ [1]

19 **Figure 10** shows a circuit.

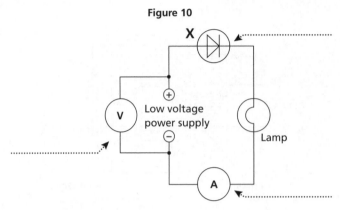

Figure 10

a) Add the missing labels to **Figure 10**. [3]

b) The circuit in **Figure 10** is tested with a range of different voltages.
The graph in **Figure 11** shows the results.

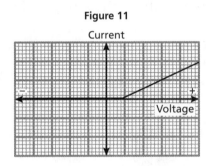

Figure 11

Describe what happens to the current through component **X** as the voltage changes.
You must refer to the resistance of the component. [4]

20 **a)** Draw a circuit diagram to show how a battery, ammeter and voltmeter can be used to find the resistance of a wire.
You must use the correct circuit symbols. [3]

b) When correctly connected to a 9V battery, the wire in the circuit has a current of 0.30A flowing through it.

i) Calculate the resistance of the wire. [2]

ii) When the wire is heated, the current drops to 0.26A.

How has the resistance of the wire changed? [1]

21 The circuit diagram in **Figure 12** includes a component labelled **X**.

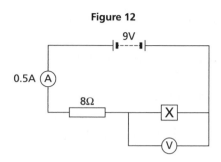

Figure 12

a) Calculate the potential difference across the 8Ω resistor.   [2]

b) What is the potential difference across component **X**?   [1]

c) What is the current through component **X**?   [1]

22 The circuit in **Figure 13** has four identical ammeters.

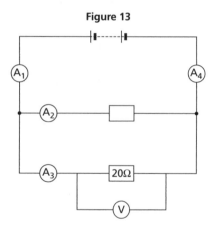

Figure 13

Table 3 gives the current through two of the ammeters.

Table 3

Ammeter	Current in amps
A$_1$	
A$_2$	0.2
A$_3$	0.3
A$_4$	

a) Complete the table to show the current through A$_1$ and A$_4$.   [2]

b) Calculate the reading on the voltmeter.
    You must show your working.   [2]

c) Give the potential difference of the power supply.   [1]

23 Microwave ovens can be used to heat many types of food.

a) Describe in detail how microwave ovens heat food. [3]

b) Microwaves used in ovens have a frequency of 2450 million Hz. Their wavelength is 0.122m.

Calculate the speed of microwaves. Show clearly how you work out your answer. [3]

24 HT **Figure 14** shows a beam of light striking a Perspex block.

Figure 14

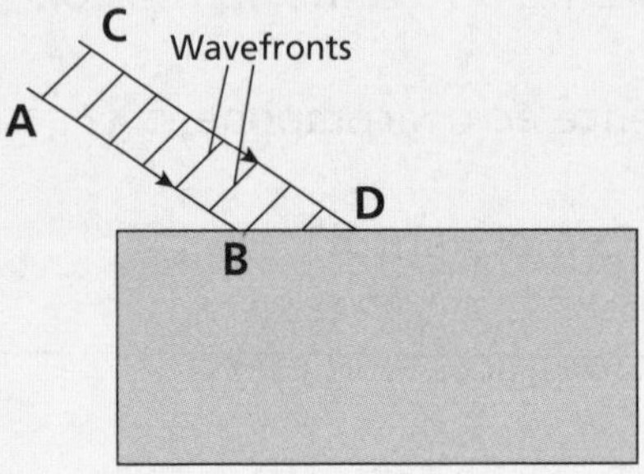

C Wavefronts

A

D

B

a) Continue the paths of the rays **AB** and **CD** inside the Perspex block on **Figure 14**. [2]

b) Draw the wavefronts of the beam of light in the Perspex. [2]

c) Explain why the beam behaves in the way you have shown. [2]

25 A student uses a ray box and a semicircular glass block to investigate refraction.

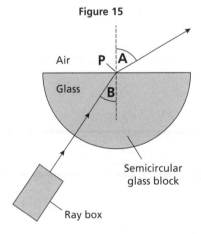

Figure 15

Air P A

Glass B

Semicircular glass block

Ray box

a) What is the vertical dashed line in **Figure 15** called? [1]

b) The student always aims the light from the ray box at point **P**.
This ensures that the ray enters at right-angles to the glass block.

Explain why this is important. [2]

The ray box is adjusted to take a range of measurements. **Table 4** shows the results.

Table 4

Angle A	Angle B
30°	19°
40°	25°
50°	31°
60°	35°
70°	39°
80°	41°

The student studies the data and comes to the conclusion that Angle **A** is directly proportional to Angle **B**.

c) Use data from the table to explain why this conclusion is **not** correct. [2]

d) Write a correct conclusion for the experiment. [2]

e) Why is your conclusion only valid when Angle **A** is between 30° and 80°? [2]

26 The arrows in **Figure 16** represent the size and direction of the forces on a space shuttle, fuel tank and booster rockets one second after launch.

Figure 16

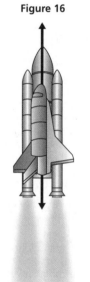

a) Name the force illustrated by the arrow pointing downwards. [1]

b) Describe the upward motion of the space shuttle one second after launch. [1]

c) By the time it moves out of the Earth's atmosphere, the total weight of the space shuttle, fuel tank and booster rockets has decreased and so has the air resistance.

How does this change the motion of the space shuttle? [2]
Assume the thrust force does not change.

d) The space shuttle takes nine minutes to reach its orbital velocity of 8100m/s.

i) Calculate the average acceleration of the space shuttle during the first nine minutes of its flight in m/s².
To gain full marks you must show how you worked out your answer. [3]

ii) How is the velocity of an object different from the speed of an object? [2]

Total Marks .... / 157

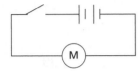

Pages 6–7 Review Questions

1. a) It moves backwards / to the left / away from the wall [1]
   b) It will slow it down [1]
2. a) A simple series circuit drawn with the correct with symbol for battery [1]; motor [1]; and switch [1]

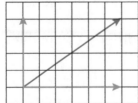

   b) The battery [1]
   c) It will be slower [1]
3. Both bulbs will go out [1]
4. Both bulbs will stay the same [1]
5. a) There is a limited amount [1]; so it will run out [1]
   b) **Any two of:** coal [1]; gas [1]; nuclear [1]
6. chemical [1]; electrical [1]; light [1]; thermal [1]
7. 30N [1]
8. a) B [1]
   b) C is bigger than A [1]
   c) B and D [1]
   d) D is bigger than B [1]
9. a) It is reflected [1]
   b) Black absorbs light / the cyclist does not reflect any light [1]; so no light from the cyclist enters the driver's eyes [1]; and they appear black against a black background [1]

Pages 8–19 Revise Questions

Page 9 Quick Test
1. A scalar quantity just has size, but a vector has size and direction.
2. Magnetism / gravity / electrostatic attraction is a non-contact force; a magnet will attract or repel another magnet without needing to touch it / one mass will gravitationally attract another / one charge will attract or repel another charge without physical contact.
3. 128N
4. 100N to the east

Page 11 Quick Test
1. 100J
2. 37.5cm

Page 13 Quick Test
1. When particles collide with the surface of an object they exert a force. The pressure is the force per unit area.
2. 10 000N
3. 900kg/m³
4. 2 000 000Pa

Page 15 Quick Test
1. Distance is the total distance travelled in any direction; displacement is the total distance from the start point and includes the direction from the start point.
2. 20m/s
3. Speed

Page 17 Quick Test
1. 2m/s²
2. The object represented by graph A is accelerating at twice the rate of the object represented by graph B.
3. To identify anomalous results, which can then be removed and allow a mean to be taken.

Page 19 Quick Test
1. 40 000kg m/s
2. As the fish swims it exerts a force backwards on the water. This creates an equal and opposite force from the water on the fish, that pushes the fish forwards.
3. a) 600 × 8 = 4800Nm
   b) $\frac{4800}{12}$ = 400kg

Page 21 Quick Test
1. It might increase the stopping distance, as it increases the thinking distance by causing a distraction.
2. Answer should be between 2160N and 3375N using city speeds as 12–15m/s.
3. One person holds a ruler with the bottom of the ruler level with the other person's hand. They let go of the ruler and the other person catches it when they see it move. The reaction time is calculated from the distance the ruler falls before being caught.
   OR
   One person presses a switch that turns on a light and starts a timer. The other person presses a switch that stops the timer when they see the light.

Pages 22–25 Practice Questions

Page 22 Forces – An Introduction
1. A scalar quantity has size only [1]; a vector quantity has size and direction [1]; mass is the measure of how much matter is in an object [1]; weight is the force of gravity and, as a force, it acts in a particular direction [1]
2. a) F_1 is twice the size of F_2 [1]; and acts in the opposite direction [1]
   b) Zero [1]
3. Correctly drawn vertical and horizontal arrows [1]

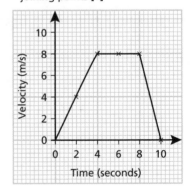

Page 22 Forces in Action
1. a) It increases [1]
   b) 150 000J [1]
   c) $\frac{150\,000}{50}$ [1]; = 3000N [1]
2. a) 60 × 10 [1]; = 600N [1]
   b) Three times the distance of the soil from the pivot, so a third of the forced needed [1]; around 200N [1] (Accept accurate calculation of 196N)

Page 23 Pressure and Pressure Differences
1. It is caused by the random movement of the particles [1]; colliding with objects or the walls of a container [1]
2. pressure = $\frac{\text{force normal to a surface}}{\text{area of that surface}}$
   / $p = \frac{F}{A}$ [1]
3. 10 × 100 [1]; = 1000N [1]

   Pressure is constant, so 10 times the area must be 10 times the force.

4. a) 100 × 1000 × 10 [1]; = 1 000 000N/m² [2] (1 mark for correct value; 1 mark for correct units)
   b) The pressure falls by [1]; 3 × 1000 × 10 [1]; = 30 000N/m² [1]
5. a) $\frac{2597}{10}$ = 259.7 [1]; = 260kg (to the nearest kg) [1]
   b) 260 + 100 = 360kg [1]
   c) 360 × 10 [1]; = 3600N [1]
   d) Mass displaced = 365kg [1]; volume = $\frac{\text{mass}}{\text{density}}$ [1]; = $\frac{365}{1000}$ [1]; = 0.365m³ [1]

Page 24 Forces and Motion
1. a) 5 miles [1]
   b) 1 mile [1]; east [1]
2. a) i) 52 × 3.65 = 189.8 miles [1]
      ii) 0 miles [1]
      iii) $\frac{189.8}{1.5}$ [1]; = 126.5mph (to 1 decimal place) [1]
   b) This occurs when direction is changing [1]; but speed is constant [1]; so, could occur on bends / corners [1]
3. a) 3 seconds [1]
   b) C to D [1]
   c) $\frac{3}{5}$ [1]; 0.6m/s [1]
   d) Change in speed / acceleration and deceleration [1]

Page 24 Forces and Acceleration
1. $v^2 = 2as + u^2$ [1]; $v^2 = (2 \times 4 \times 100) + 10^2$ [1]; $v^2 = 900$ [1]; $v = \sqrt{900}$ = 30m/s [1]
2. a) Correct axes and labels [1]; accurately plotted points [1]; straight lines joining points [1]

b) A constant speed of 8m/s **[1]**

c) $\frac{8}{4}$ **[1]**; = 2m/s² **[1]**

d) Area under graph divided into three sections **[1]**;
= $\frac{(8 \times 4)}{2} + (8 \times 4) + \frac{(8 \times 2)}{2}$ **[1]**;
= 56m **[1]**

Page 25 Terminal Velocity and Momentum

1. acceleration = $\frac{\text{change in velocity}}{\text{time}}$ /
$a = \frac{\Delta v}{t}$ **[1]**; $a = \frac{(4-0)}{0.4}$ **[1]**; = 10m/s² **[1]**

2. a) 25 000N **[1]**
   b) The helicopter pushes the air downwards **[1]**; which creates an equal and opposite force from the air on the helicopter, which pushes the helicopter upwards **[1]**; when the force exerted on the air downwards is equal to the weight of the helicopter, the helicopter hovers at a constant height **[1]**

Page 25 Stopping and Braking

1. force = rate of change of momentum **[1]**; during a crash, the driver experiences a change of momentum and the faster this change happens, the bigger the force and the more serious the injury **[1]**; the airbag deflating increases the time that change takes place over **[1]**; and, therefore, reduces the force on the driver **[1]**

2. **Any three of:** alcohol **[1]**; fatigue **[1]**; drugs **[1]**; distractions **[1]** (Accept specific distractions, e.g. mobile phone use)

3. $\frac{5000 \times 12}{5}$ **[1]**; = 12 000N **[1]**

Pages 26–41 Revise Questions

Page 27 Quick Test

1. Diagram should show chemical to gravitational (+ wasted heat)
2. 504 000J
3. Reproducible means that another person could follow your method and get the same answer. It is important because it helps to show that the results are valid.

Page 29 Quick Test

1. It is released as heat and lost to the surroundings.
2. To build it you need to flood valleys and displace inhabitants / destroy habitats.

Page 31 Quick Test

1. At right-angles to the direction of energy transfer.
2. 0.05s
3. If the waves are visible, count the number of waves passing a fixed point in a second. Measure the wavelength. Multiply these two measurements together.

Page 33 Quick Test

1. A line drawn at right-angles to the surface at the point of origin.
2. Away from the normal.
3. 20 000Hz

Page 35 Quick Test

1. 900m
2. P-waves are faster than S-waves; P-waves are longitudinal, S-waves are transverse.
3. S-waves are not able to travel through the liquid outer core of the Earth, so there is a shadow formed on the other side of the Earth, where these waves are not detected when there is an earthquake.

Page 37 Quick Test

1. Radio waves, microwaves, infrared waves, visible light, ultraviolet waves, X-rays, gamma rays
2. Microwaves are absorbed by food and heat it up. (Accept any other sensible answer)
3. They are easily absorbed by water.

Page 39 Quick Test

1. A real image can be projected onto a screen; a virtual image cannot.
2. A convex / converging lens
3. 3cm

Page 41 Quick Test

1. Light passes through a transparent object coherently so you can see through it clearly; a translucent object scatters light, so objects cannot be seen clearly.
2. A green object only reflects green light. A red filter only lets red light through it. So, no light comes through the filter and the object looks black.
3. Magnesium burns at a higher temperature.

Pages 42–47 Review Questions

Page 42 Forces – An Introduction

1. a) Short arrow drawn in direction of friction **[1]**

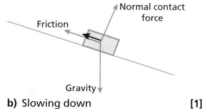

   b) Slowing down **[1]**
2. A contact force only occurs where objects are touching **[1]**; a non-contact force can be felt at a distance **[1]**
3. gravity **[1]**; magnetism **[1]**; non-contact **[1]**; drag **[1]**; contact **[1]** (Accept gravity / magnetism in any order)
4. a) i) 70 × 1.6 = 112N **[1]**
      ii) 70 × 10 = 700N **[1]**
   b) The force from the astronaut's legs, and the speed they would leave the ground at, is unchanged **[1]**; but

the force acting downwards is much less, so the astronaut travels higher before coming back down **[1]**

5. Correctly drawn driving force **[1]**; frictional force **[1]**; and resultant force **[1]**; resultant force of 15N stated **[1]**

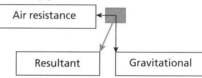

Frictional 5N ← 🔲 → Driving 20N

Resultant 15N

6. Correctly drawn air resistance **[1]**; gravitational force **[1]**; and resultant force **[1]**

Air resistance	
Resultant	Gravitational

Page 43 Forces in Action

1. work done = force × distance (along the line of action of the force) / $W = Fs$ **[1]**
2. Five floors = 5 × 3.5 = 17.5m **[1]**; work done = 1200 × 17.5 **[1]**; = 21 000J **[1]**
3. Weight = 40 × 10 = 400N **[1]**; work done = 400 × 3 **[1]**; = 1200J **[1]**
4. There needs to be two forces pulling in opposite directions **[1]**; or the spring would just move **[1]**
5. The limit of proportionality is the point that once exceeded **[1]**; the extension is no longer directly proportional to the applied force **[1]**
6. The fish exerts a moment of 12 × 4 = 48Nm **[1]**; for the rod to remain still, the fisherman must exert an equal and opposite moment **[1]**; he must apply a force of $\frac{48}{1}$ **[1]**; = 48N **[1]**
7. The rubber band obeys Hooke's law up to a force of 5N **[1]**; after this it has passed the limit of proportionality **[1]**; and the amount of force required to keep increasing the length increases **[1]**

Page 43 Pressure and Pressure Differences

1. a) force normal to a surface = pressure × area of surface / $F = pA$ **[1]**; 100 000 × 10 = 1 000 000N **[1]**
   b) The weight of air is felt as air pressure **[1]**; the pressure under the roof is the same as above but pushes up on the roof **[1]**; balancing the pressure pushing down on the roof **[1]**
2. Force = 200 × 5 **[1]**; = 1000N **[1]**

   The area is five times greater, so multiply by 5.

3. a) Height = 0.1m **[1]**; pressure = 0.1 × 13 500 × 10 **[1]**; = 13 500N/m² **[1]**
   b) force normal to a surface = pressure × area of that surface / $F = p \times A$ **[1]**; = 13 500 × 0.0003 **[1]**; = 4.05kg **[1]**

   You need to use the equation $p = \frac{F}{A}$ and rearrange it to find the force. Don't forget to change the area from cm² to m²: 1cm² = 0.0001m².

c) The pound coin is less dense than mercury [1]

d) 0.1N [1]

4. There are few particles [1]; to collide with objects creating pressure [1]; and the height of air above is less [1]

Page 44 Forces and Motion

1. Velocity is a vector so has direction and magnitude [1]; speed has only magnitude [1]

2. distance travelled = speed × time / $s = vt$ [1]

3. a) 3 + 2 + 1 = 6 miles [1]

 b) Correctly drawn vector diagram with three arrows showing the stages of the journey [1]; and one arrow showing the displacement [1]

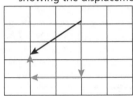

4. $\frac{4500}{3}$ [1]; = 1500m/s [1]

5. a) High speed / velocity [1]

 b) Stationary / at rest [1]

 c) Changing speed / acceleration or deceleration [1]

6. A sketch showing correctly labelled axes [1]; and a graph line with a gentle upward curve at the start [1]; followed by a straight, upward slope [1]; followed by a curved slope with a shallower gradient [1]

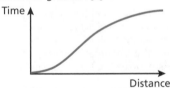

7. A planet in orbit / a roundabout / a car on a bend (any other sensible example) [1]; it is changing direction so the velocity changes [1]; but the magnitude of velocity i.e. the speed is constant [1]

8. Inertia [1]

9. a) Resultant force = 0 [1]; at rest [1]

 b) Resultant force = 0 [1]; constant speed [1]

Page 45 Forces and Acceleration

1. a) True [1]

 b) False [1]

 c) False [1]

2. $0 - 20^2 = (2 \times -2 \times s)$ [1]; $s = \frac{-400}{-4}$ [1]; $s = 100m$ [1]

3. a) Speeding up [1]

 b) Slowing down [1]

 c) Constant forwards / positive velocity [1]

 d) Constant backwards / negative velocity [1]

 e) Distance travelled [1]

Page 46 Terminal Velocity and Momentum

1. a) 4 [1]

 b) 5 [1]

 c) 1 [1]

 d) 4 / 7 [1]

2. The blades on the propeller push the air backwards [1]; this creates an equal and opposite force of the air pushing on the hovercraft [1]; which moves it forwards [1]

3. momentum = mass × velocity / $p = mv$ [1]

4. Momentum of the cannon ball = 1 × 120 = 120kgm/s [1]; since momentum is conserved, the momentum of the cannon is 120kgm/s in the opposite direction [1]; $v = \frac{p}{m} = \frac{120}{240}$ = [1]; = 0.5m/s [1]

5. The yellow ball has $\frac{3}{4}$ of the initial momentum of the cue ball [1]; since the balls have the same mass, the velocity of the yellow ball = $\frac{3}{4} \times 4$ = 3m/s [1]

Page 47 Stopping and Braking

1. **Any three of:** rain / snow [1]; old / worn tyres [1]; old / worn brakes [1]; poor road condition [1]

2. He is incorrect [1]; doubling the speed quadruples the kinetic energy [1]; so the brakes need to do four times the work and it takes four times distance to stop [1]

3. Brakes use friction to slow the car [1]; this does work on the car by converting kinetic energy [1]; to heat energy [1]; on a long steep slope the brakes are more likely to overheat causing them to fail [1]

4. The reaction time is the time it takes a person to react to something [1]; the thinking distance is the distance a vehicle travels during this reaction time before the driver applies the brakes [1]; the stopping distance is the sum of thinking and braking distance [1]

5. Thinking distance = 12 × 0.5 = 6m [1]; braking distance = 6 + 14 [1]; = 20m [1]

6. a) True [1]

 b) True [1]

 c) True [1]

 d) True [1]

Page 48 Energy Stores and Transfers

1. a) i) Chemical Energy [1]

 ii) Kinetic Energy [1]

 iii) Sound Energy [1]

 iv) Heat Energy [1]

 b) Each square represents 100kJ [1]; so, kinetic energy = 300kJ [1]

2. a) She begins with gravitational energy [1]; as she descends this is converted into kinetic energy [1]; once she reaches the bottom, she has the maximum amount of kinetic energy and the minimum amount of gravitational energy [1]; as she travels up the other side, kinetic energy

is converted back to gravitational energy [1]; during this some energy is converted to sound [1]; and heat, which are lost to the surroundings [1]

 b) kinetic energy = 0.5 × mass × (speed)² / $E_k = \frac{1}{2}mv^2$ [1]; = 0.5 × 60 × 4² [1]; = 480J [1]

 c) gravitational potential energy = mass × gravitational field strength × height / $E_p = mgh$ [1]; = 60 × 10 × 2 [1]; = 1200J [1]

3. Extension = 0.1m [1]; 0.5 × 100 × 0.1² [1]; = 0.5J [1]

Page 48 Energy Transfers and Resources

1. It will eventually run out and cannot be replaced [1]

2. It depends on the thermal conductivity of the materials [1]; the thickness of the materials [1]; and the difference in temperature between the inside and outside [1]

Page 49 Waves and Wave Properties

1. When a stone is dropped in water the resulting waves move outwards [1]; but no hole is left in the middle [1] (Accept any other sensible example)

2. particles [1]; oscillate [1]; fixed [1]; energy [1]

3. a) Correct labels showing amplitude [1]; and wavelength [1] (Accept an arrow for amplitude from the midpoint to the maximum negative displacement and any arrow for wavelength beginning and ending at an equivalent position of the repeating wave)

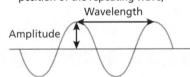

 b) 2 [1]

 c) 2 × 2 = 4 seconds [1]

4. The wavelength would halve [1]

5. a) frequency = $\frac{1}{period}$ [1]; $\frac{1}{0.05}$ = 20Hz [1]

 b) wavelength = $\frac{wave\ speed}{frequency}$ / $\lambda = \frac{v}{f}$ [1]; $\frac{330}{20}$ = 16.5m [1]

6. Stand a known distance from a large wall or cliff face and bang two pieces of wood together [1]; measure the time between banging the wood and hearing the echo [1]; divide this time by two [1]; and then divide into the distance to the wall to give the speed of sound [1]

Page 50 Reflection, Refraction and Sound

1. Correct drawing showing correctly placed mirror [1]; horizontal ray line [1]; vertical ray line [1]; and arrow heads on both ray lines [1]

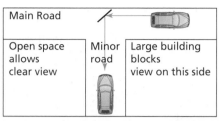

Main Road		
Open space allows clear view	Minor road	Large building blocks view on this side

2. Correct drawing showing normal where light enters the block **[1]**; ray line bending towards normal **[1]**; normal where light exits the block **[1]**; ray line bending away from normal **[1]**

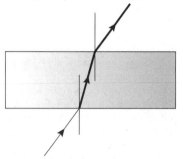

3. Its speed changes **[1]**
4. It is faster in a vacuum **[1]**
5. a) Amplitude **[1]**
   b) Frequency **[1]**

Page 50 Waves for Detection and Exploration

1. The receiver in the scanner would detect the reflected pulse sooner **[1]**; indicating that the total distance travelled is less **[1]**
2. Using $s = vt$ **[1]**; the total distance travelled by the pulse is 1500 × 0.1 **[1]**; = 150m **[1]**; $\frac{150}{2}$ = 75m **[1]**

   > Don't forget to halve the total distance to find the distance to the shoal.

3. a) Primary / P-waves **[1]**; secondary waves / S-waves **[1]**
   b) The time they arrive at different sensors can be used to determine the direction of the earthquake **[1]**; and the distance between P-waves and S-waves can be used to calculate how far away it is **[1]**
4. a) A section of the Earth where seismic waves from an earthquake are not detected **[1]**
   b) That there is a liquid outer core **[1]**; and a solid inner core **[1]**

Page 51 The Electromagnetic Spectrum

1. a) Microwaves **[1]**
   b) An opinion **[1]**
   c) i) Short to medium term use by adults does not cause brain tumours **[1]**
      ii) Microwaves are not ionising radiation **[1]**; they are low frequency **[1]**; and do not carry enough energy to cause the DNA of a cell to mutate when they are absorbed **[1]**

Page 52 Lenses

1. It is wider in the middle **[1]**; than at the edges **[1]** OR it is convex **[2]**
2. Correctly drawn diagram showing ray lines meeting at a point **[1]**; labelled 'focus' or 'focal point' **[1]**

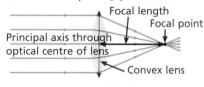

3. The distance between the lens and the points where parallel rays of light entering the lens are brought to a focus **[1]**
4. distant **[1]**; parallel **[1]**; focal point **[1]**; principal axis **[1]**
5. Correctly drawn diagram showing ray lines **[1]**; and image **[1]**

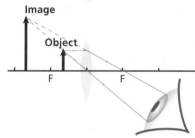

Page 53 Light and Black Body Radiation

1. Specular reflection **[1]**; where the reflection is in a single direction, like a mirror **[1]**; and diffuse reflection **[1]**; where the rays are scattered, like the reflection from a piece of paper **[1]**
2. Three correctly drawn lines **[2]** (1 mark for one line)
   Transparent – Light passing through this remains coherent, so objects can be seen clearly
   Translucent – Allows light to pass through, but light rays are 'jumbled' so objects are obscured
   Opaque – Does not let any light pass through
3. A blue car will only reflect blue light **[1]**; so when lit by white light, it reflects the blue light from within the white light **[1]**; but when lit by an orange light, there is no blue to reflect **[1]**; so it absorbs all the light and appears black **[1]**
4. a) The higher the temperature, the faster it emits radiation **[1]**
   b) As it gets continually hotter, it begins to glow a dull red **[1]**; then a bright red **[1]**; and eventually white **[1]**
5. a) As the sun comes up **[1]**; the car absorbs more energy than it emits **[1]**
   b) As the car heats up, it emits more energy **[1]**; and eventually reaches a balance between energy absorbed and energy emitted **[1]**
   c) The sun could have gone behind a cloud, reducing the energy

absorbed **[1]**; or a door / window could have been opened, increasing the energy emitted **[1]**

Page 55 Quick Test

1. 10A
2.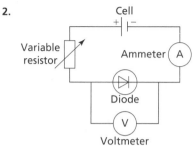
3. 5Ω

Page 57 Quick Test

1. Diode
2. Thermistor

Page 59 Quick Test

1. The resistance of R_2 is one-fifth that of R_1; it has a 1V potential difference across it
2. $P = I^2R = 2^2 \times 6 = 24$W
3. $P = IV$, $P = 30 \times 230 = 6900$W

Page 61 Quick Test

1. An alternating current changes direction rapidly and goes backwards and forwards; a direct current travels in one direction only.
2. 230V
3. 3W, 30J
4. 100 000J

Page 63 Quick Test

1. 2.5V
2. A Sankey diagram with an electrical input showing a 50/50 split between useful energy (heat and kinetic) and waste energy (heat and sound).

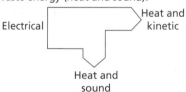

3. 3 600 000J
4. The power is generated in large stations; the electricity is transferred at high voltage / low current.

Page 65 Quick Test

1. A metal slide would earth the charge, so a charge could not build up.
2. They will repel each other.
3. The arrows around the large positive

charge will point away from the charge and there will be many field lines. The small negative charge will have fewer field lines and the arrows will point inwards.

4. As it moves further away, the field lines are further apart so the field is weaker and the force is weaker.

Page 67 Quick Test
1. The magnet is very strong.
2.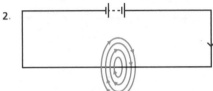
3. Increase current / more turns of wire in the coil.

Page 69 Quick Test
1. Increase current / use larger magnets
2. The coil would not be free to spin and the current would not stay in the same direction, so the coil would move to one position and stay there.
3. Higher frequency input of alternating current
4. To make the speaker move back and forward, otherwise it would move once and stop moving instead of vibrating.

Page 71 Quick Test
1. Increase the speed at which the magnet or coil move.
2. It would induce a higher potential difference.
3. A step-up transformer has more turns on the secondary coil than the primary and the output voltage is higher than the input; a step-down transformer is the opposite.

Pages 72–79 Review Questions

Page 72 Energy Stores and Transfers
1. a) Work done by motor = the gravitational potential energy [1]; $2200 \times 10 \times 16.2$ [1]; = 356 400J [1]
   b) 356 400J [1]
   c) $v^2 = \dfrac{2 \times 356\,400}{2200} = 324$ [1];
   $v = \sqrt{324} = 18\text{m/s}$ [1]
2. a) Extension = 0.6m [1]; energy stored = $0.5 \times 100 \times 0.6^2$ [1]; =18J [1]
   b) $E_k = 18\text{J} = 0.5 \times 0.0576 \times v^2$ [1];
   $v^2 = \dfrac{18}{(0.5 \times 0.0576)} = 625$ [1]; $v = \sqrt{625}$
   = 25m/s [1]
   c) Energy is transferred to the surroundings [1]; as heat and sound [1]

Page 72 Energy Transfers and Resources
1. 1 mark for each correctly completed row [4]

Input Energy	Situation	Useful Output Energy
Chemical / Kinetic	Walking uphill	Gravitational
Gravitational	Sliding down a slide	Kinetic
Elastic	Firing a catapult	Kinetic
Chemical	Fireworks / sparklers	Light and sound

2. a) Once the panels are installed [1]; they will get energy from the Sun, which does not cost anything [1]
   b) You have to pay for the panels to be installed [1]; and maintained [1]

Page 73 Waves and Wave Properties
1. wave speed = frequency × wavelength / $v = f\lambda$ [1]
2. a) i) Transverse [1]
      ii) Move their hand up and down faster [1]
      iii) Move the end of the spring up and down through a greater distance [1]
   b) i) Longitudinal [1]
      ii) Sound [1] (Accept any other sensible answer)
   c) Energy [1]

Page 74 Reflection, Refraction and Sound
1. Correctly drawn diagram with lines extending beyond mirror [1]; and image drawn [1]; and line reflecting off mirror [1]; with object drawn [1]

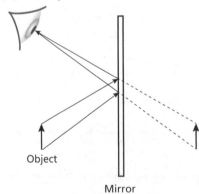

Object

Mirror

2. As waves move from one medium to another, they change speed [1]; if entering at an angle, the part that crosses the boundary first changes speed first [1]; if it slows down, the wave will turn towards the normal [1]; if it speeds up, the wave will turn away from the normal [1]
3. The refractive index of the block should be controlled [1]; by using the same block [1]; if the block is changed, the amount of refraction could change, even when the angle is the same [1]

4. As the waves passes through the solid, oscillations in the wave [1]; cause the particles in the solid to oscillate [1]; with the same frequency as the wave [1]
5. Ear drum [1]; microphone [1] (Accept any other sensible answer)

Page 74 Waves for Detection and Exploration
1. The ultrasound is transmitted into the patient and when it meets a boundary between materials of different density [1]; it is partially reflected [1]; the reflected wave from a foetus is detected, indicating its presence [1]
2. a) They are both produced by earthquakes [1]
   b) P-waves are longitudinal and S-waves are transverse [1]; P-waves travel faster [1]
3. Station A is very close to the epicentre [1]; Station B is a long distance away [1]; but on the same side of the planet as the epicentre [1]; Station C is on the opposite side of the planet to the epicentre [1]
4. Correctly drawn diagram showing waves reflecting off the right side of the outer core [1]; and left side of the outer core [1]; with a shadow zone drawn and labelled on the underside of the planet [1]

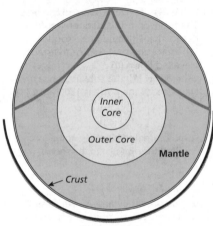

Shadow zone

5. distance = speed × time / $s = vt$
   = $1500 \times 0.04 = 60\text{m}$ [1]; depth = $\dfrac{60}{2}$ [1];
   = 30m [1]

Page 75 The Electromagnetic Spectrum
1. a) Visible light is absorbed by the plant [1]; providing it with the energy it needs to grow [1]
   b) Infrared waves are absorbed by the bread [1]; heating it up and toasting it [1]
   c) Radio waves are absorbed by an aerial [1]; creating an electrical signal in the aerial [1]
2. high [1]; most [1]; bacteria [1]
3. Ionising radiations [1]; can knock electrons from atoms ionising them [1]; when ionising radiation is absorbed by the nucleus of a cell [1]; the ionisation

can cause DNA to mutate and cells to become cancerous **[1]**
4. Alternating current **[1]**

Page 76 Lenses

1. a) Correctly drawn diagram showing one straight, sloping ray **[1]**; one horizontal ray being refracted **[1]**; and the image drawn at the point where the two cross **[1]**

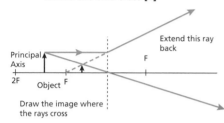

b) On top diagram: two correctly drawn rays **[2]**; and a correctly drawn image **[1]**
On bottom diagram: two correctly drawn rays **[2]**; and a correctly drawn image **[1]**

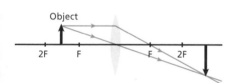

2. A real image can be projected onto a screen, but a virtual image cannot **[1]**; a real image is on the opposite side of the lens to the object and a virtual image is on the same side **[1]**; you need to look through the lens to see a virtual image **[1]**

Page 76 Light and Black Body Radiation

1. It can be absorbed **[1]**; reflected **[1]**; or transmitted **[1]**
2. a) The toy reflects red light from the red stripes **[1]**; and blue light from the blue stripes **[1]**; all other colours are absorbed **[1]**
   b) Blue **[1]**; and black **[1]**
3. A black object reflects no visible light **[1]**; a white object reflects all wavelengths of light **[1]**; equally **[1]**
4. It will absorb all the electromagnetic radiation incident on it **[1]**; without reflecting any **[1]**
5. a) The star is a higher temperature **[1]**; so the wavelengths of light it emits are throughout the visible spectrum **[1]**; the hob only

emits light in the red part of the spectrum **[1]**
   b) An object is hot enough to cause burns at 100°C **[1]**; but it will not give out visible wavelengths until well past this temperature **[1]**; at this temperature only infrared is emitted **[1]**
   c) 10 000K **[1]**

Pages 78–83 Practice Questions

Page 78 An Introduction to Electricity

1. Diagram showing a correctly connected battery **[1]**; bulb **[1]**; ammeter **[1]**; and voltmeter **[1]**

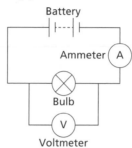

2. a) Cell **[1]**
   b) Variable resistor **[1]**
   c) LED **[1]**
   d) Thermistor **[1]**

3. a) resistance = $\frac{\text{potential difference}}{\text{current}}$ /
      $R = \frac{V}{I}$ **[1]**; = $\frac{12}{0.5}$ **[1]**; = 24Ω **[1]**
   b) Twice the current will flow **[1]**
4. a) The variable resistor allows a larger number of measurements **[1]**; with smaller intervals, more precise results can be recorded **[1]**
   b) This prevents the circuit from becoming hot **[1]**; which could affect resistance **[1]**
   c) Resistance (R) = 50Ω in darkness and 1Ω in maximum brightness **[1]**; maximum difference = 50 − 1 = 49Ω **[1]**

Page 79 Circuits and Resistance

1. a) It is constant **[1]**
   b) A line drawn that passes through zero **[1]**; but is less steep than the first line **[1]**
2. Zero **[1]**

Page 79 Circuits and Power

1. a) Diagram showing a correctly connected 6V battery **[1]**; and three bulbs in series **[1]**

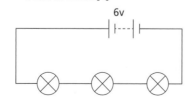

b) $\frac{6}{3}$ = 2V **[1]**
   c) 2A **[1]**
   d) resistance = $\frac{\text{potential difference}}{\text{current}}$ /
      $R = \frac{V}{I}$ **[1]**; = $\frac{2}{2}$ **[1]**; = 1Ω **[1]**
   e) 3Ω **[1]**
2. a) Increases the total resistance **[1]**
   b) Decreases / reduces the total resistance **[1]**

Page 80 Domestic Uses of Electricity

1. Live – brown **[1]**; neutral – blue **[1]**; earth – yellow and green **[1]**
2. The live wire is at a high potential **[1]**; and a person is at zero **[1]**; touching the wire creates a large potential difference **[1]**; so a large current flows to earth through the person **[1]**
3. To complete the circuit **[1]**
4. 0V **[1]**

Page 80 Electrical Energy in Devices

1. a) The amount of energy transferred per second **[1]**
   b) power = potential difference × current / $P = VI$ **[1]**; $I = \frac{1000}{230}$ **[1]**; = 4.35A **[1]**
2. power = (current)² × resistance / $P = I^2R$ **[1]**
3. a) power = (current)² × resistance / $P = I^2R$ **[1]**; $P = 0.1^2 × 5000$ **[1]**; = 50W **[1]**
   b) energy transferred = power × time / $E = Pt$ **[1]**; $E = 50 × 2 × 60$ **[1]**; = 6000J **[1]**
4. a) TV **[1]**
   b) Comparing power × time for each appliance **[1]**; the tumble dryer has the greatest value **[1]**
   c) Time (t) = 1 × 60 × 60 = 3600s **[1]**; energy transferred = power × time / $E = Pt$ **[1]**; $E = 5000 × 3600$ **[1]**; = 18 000 000J **[1]**
   d) charge = $\frac{\text{energy transferred}}{\text{potential difference}}$ / $Q = \frac{E}{V}$ **[1]**; $Q = \frac{18\,000\,000}{240}$ = 75 000C **[1]**
5. It keeps the heating effect of the current low **[1]**; reducing wasted energy **[1]**

Page 81 Static Electricity

1. a) Electrons **[1]**; have been transferred from the jumper **[1]**; to the balloon **[1]**

Answers

b) The paper would be attracted to the balloon **[1]**

2. **a)** All of the hairs become charged **[1]**; with the same type of charge **[1]**; so they are repelled from the head and each other **[1]**

   b) There is a large potential in the static charge on the generator **[1]**; and the rod is at earth **[1]**; so there is a large potential difference **[1]**; which, when large enough, causes a current to flow through the air as a spark **[1]**

3. The direction a positive charge will move in if placed in the field **[1]**

Page 82 Magnetism and Electromagnetism

1. Correctly drawn diagram showing lines from north to south pole **[1]**; coming out of the north pole into the south pole **[1]**; more widely spaced the further they are from the magnet **[1]**

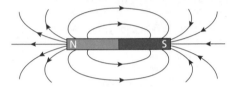

2. When a material becomes a magnet due to the presence of a permanent magnet or an electric current **[1]**

3. **a)** A bigger current **[1]**; putting the compass closer to the wire **[1]**

   b) By reversing the current **[1]**

   c) A downward pointing arrow drawn on the diagram **[1]**

Page 82 The Motor Effect

1. **a)** Fleming's left hand rule **[1]**

   b) Upwards **[1]**

   c) Motors **[1]**; loudspeakers **[1]** (Accept any other sensible answer)

   d) Bigger current / stronger magnets **[1]**

   e) Reverse the current / reverse the magnets **[1]**

2. Force = $(50 \times 10^{-6}) \times 12 \times 100$ **[1]**; = 0.06N **[1]**

3. **a)** Force = $0.3 \times 1 \times 0.02$ **[1]**; = 0.006N **[1]**

   b) 1.2N **[1]**

Page 83 Induced Potential and Transformers

1. **a)** **i)** iron core **[1]**

      ii) primary coil **[1]**

      iii) secondary coil **[1]**

   b) Step-down **[1]**

   c) Voltage ratio = $\frac{230}{12}$ **[1]**; turns on coil = $4600 \times \frac{12}{230}$ **[1]**; = 240 turns **[1]**

Pages 84–97 Revise Questions

Page 85 Quick Test

1. Because there are no gaps between the particles.

2. By putting the object in a liquid. The amount of displaced liquid is equal to

---

the volume. Then find the object's mass and use the equation density = $\frac{mass}{volume}$ to calculate the density.

3. The particles would move faster and hit surfaces with a bigger force and more frequently; the pressure would increase because $P = \frac{F}{A}$.

Page 87 Quick Test

1. Positive / +1

2. An element is defined by the number of protons it has. Isotopes of an element have the same number of protons but different numbers of neutrons.

3. The mass number tells us the total number of protons plus neutrons in the nucleus.

4. **a)** That some alpha particles were reflected.

   b) This showed that most of the mass of an atom is in one place and has a positive charge.

Page 89 Quick Test

1. Gamma rays

2. Because beta has a longer range in air and can penetrate the skin to damage internal organs; alpha has a very short range in air and is unlikely to damage internal organs unless the radioactive source is inside the body.

3. Contamination leaves the substance radioactive but irradiation does not.

Page 91 Quick Test

1. It is not possible to predict exactly when it will occur.

2. 2 months

3. $\frac{1}{32}$

Page 93 Quick Test

1. Natural: rocks; man-made: X-rays (Accept any other sensible answer)

2. They are absorbed by other nuclei causing them to undergo fission.

Page 95 Quick Test

1. Protostar

2. moon, planet, star, galaxy, universe

3. The expansive forces caused by fusion and the force of gravity.

4. Through fusion, by fusing lighter nuclei into heavier ones.

Page 97 Quick Test

1. It is constant.

2. Gravity

3. The wavelength gets longer, so appears redder.

4. That they are all moving away from each other.

Pages 98–103 Review Questions

Page 98 An Introduction to Electricity

1. **a)** 1 hour = 3600s **[1]**; $Q = 0.1 \times 3600$ **[1]**; = 360C **[1]**

   b) Twice the voltage would be needed **[1]**

---

2. current **[1]**; voltage **[1]**; current **[1]**; insulator **[1]**

3. **a)** 0.1A **[1]**

   b) 20 ohms **[1]**

4. potential difference = current × resistance / $V = IR$ **[1]**; $V = 2 \times 12 = 24V$ **[1]**

Page 98 Circuits and Resistance

1. **a)** **i)** Graph 3 **[1]**

      ii) Graph 1 **[1]**

      iii) Graph 2 **[1]**

   b) Graph 1 **[1]**

   c) Graph 2 **[1]**

Page 99 Circuits and Power

1. **a)** Diagram showing correctly connected 2V battery **[1]**; and two bulbs in series **[2]** (1 mark for each correctly drawn pathway and bulb)

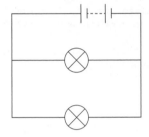

   b) 2V **[1]**

   c) 1A **[1]**

   d) resistance = $\frac{potential\ difference}{current}$ / $R = \frac{V}{I}$ **[1]**; = $\frac{2}{1}$ = 2Ω **[1]**

Page 99 Domestic Uses of Electricity

1. 230V **[1]**; 50Hz **[1]**

2. **a)** Correctly labelled earth wire **[1]**

   b) Correctly labelled live wire **[1]**

   c) Correctly labelled neutral wire **[1]**

   d) Correctly labelled fuse **[1]**

   e) Correctly labelled cable grip **[1]**

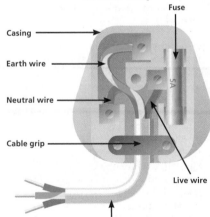

The Three-Pin Plug

Page 100 Electrical Energy in Devices

1. power = potential difference × current / $P = VI$ **[1]**

2. a) current = $\dfrac{\text{power}}{\text{potential difference}}$

   / $I = \dfrac{P}{V}$ [1]; $I = \dfrac{2600}{230}$ [1]; = 11.3A [1]

   b) energy transferred = power × time /
   $E = Pt$ [1]; $E = 2600 \times 30 \times 60$ [1];
   = 4 680 000J [1]

   c) charge flow = current × time / $Q = It$ [1];
   $Q = 11.3 \times 30 \times 60$ [1]; = 20 340C [1]

3. a) i) Step-up transformer   [1]
      ii) Step-down transformer   [1]
   b) It reduces the current [1]; to reduce
   energy loss on transmission [1]
   c) It reduces the voltage [1]; to safe
   levels for consumers to use [1]

4. It saves resources [1]; reduces
   environmental damage [1]; saves
   money [1]

5. The efficiency of steam turbines
   depends on the temperature that
   the steam can reach [1]; large power
   stations can reach higher temperatures
   [1]; and transporting fuel to the station
   can be more cost effective by using
   large trains [1]

Page 101 Static Electricity

1. a) It is isolated from the earth [1]; by an
   insulator [1]
   b) It would conduct the charge to
   earth [1]

2. This makes sure there is no charge built
   up on the lorry [1]; which could make a
   spark [1]; and cause the fuel to ignite [1]

3. On first diagram: lines with arrows from
   positive to negative [1]; lines further apart
   as they get further away from centre [1]
   On second diagram: lines with arrows
   all pointing outwards [1]; lines further
   apart as they get further away from
   centre [1]

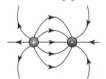

Page 101 Magnetism and Electromagnetism

1. a) It concentrates the field from the
   whole wire [1]; into a smaller
   area [1]
   b) Increasing the voltage [1]; or
   current [1]
   c) An iron core [1]

2. Diagram labelled with north pole [1];
   with field lines with arrows running from
   north to south [1]; lines further apart as
   they get further away from centre [1]

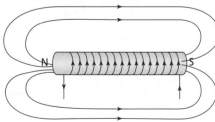

3. Around the poles   [1]
4. Magnet A is stronger than Magnet B   [1]

Page 102 The Motor Effect

1. a) An upward pointing arrow   [1]

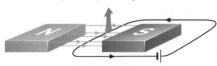

   b) It will reverse it   [1]
   c) The wire will oscillate backwards
   and forwards [1]; at a frequency of
   50Hz [1]

2. a) How strong the magnetic field is [1];
   shown by how close together the
   field lines are [1]
   b) Teslas   [1]

3. magnetic flux density =

   $\dfrac{\text{force on conductor}}{\text{current } \times \text{ length of wire}}$ / $B = \dfrac{F}{Il}$ [1];

   magnetic flux density = $\dfrac{0.2}{4 \times 0.04}$ [1];

   = 1.25T [1]

   > Don't forget to convert length
   > from centimetres to metres before
   > performing the calculation.

Page 103 Induced Potential and Transformers

1. a) Sketch with correctly labelled axes
   [1]; and correct shape of graph [1]

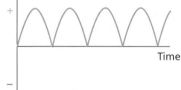

   Voltage V

   b) When the coil is at right-angles to the
   field, the induced potential is zero [1];
   the peak voltage occurs when the coil
   is parallel to the field [1]; because the
   rotating coil cuts the field lines the
   quickest at this point [1]

2. a) The incoming sound wave makes the
   cone vibrate [1]; this then makes the
   coil move relative to the magnetic
   field [1]; which generates a current
   in the coil with the same frequency
   as the movement of the cone [1]
   b) The cone would move with a greater
   distance, so the signal would have
   greater amplitude   [1]

3. A transformer works because the input
   alternating current induces a changing
   magnetic field in the core [1]; which
   induces a changing current in the
   secondary coil [1]; if the input was a
   direct current, the core would become
   an electromagnet [1]; but the field
   would not change, so no current would
   be induced in the secondary coil [1]

Page 104 Particle Model of Matter

1. Cannot be squashed   [1]
2. An object less dense will float on the
   water and will not be submerged [1]; so
   the volume measured would be too low
   [1]
3. The input energy is being used to give
   the particles enough energy to escape
   from the liquid [1]; as the water becomes
   a gas, the particles move much faster [1]
4. When the steam condenses into water,
   it releases the energy it needed to be a
   gas [1]; so steam contains more energy
   than the same mass of water at the
   same temperature [1]
5. As the temperature increases, the
   particles move faster and hit the wall
   of the container more often/with more
   energy [1]; this increases the pressure
   [1]; and the force that the particles
   exert on the lid [1]
6. Pressure × volume remains constant [1];
   since the volume decreases by a

   factor of $\dfrac{40}{50}$ [1]; the pressure

   increases to $100 000 \times \dfrac{50}{40}$ [1];
   = 125 000Pa [1]
7. Object A – No [1]; Object B – Yes [1];
   Object C – Yes [1]

   > Remember, density = $\dfrac{\text{mass}}{\text{volume}}$

Page 105 Atoms and Isotopes

1. a) Diagram correctly labelled with
   protons [1]; neutrons [1]; and
   electrons [1]

   Helium Atom

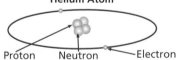

   Proton   Neutron   Electron

   b) Positive / +1   [1]
   c) 1×10^{-10}m   [1]
   d) It would gain electrons   [1]
   e) Mass number 4 [1]; atomic number
   2 [1]

2. 6   [1]
3. a) 17   [1]
   b) 17   [1]
   c) Chlorine-37 has 20 neutrons [1];
   chlorine-35 has 18 neutrons [1]
4. electrons [1]; nucleus [1]; radiation [1];
   electrons [1]; higher [1]; lower [1]
5. A ball of positive charge [1]; with
   negative electrons embedded in it [1];
   correctly drawn diagram [1]

Answers

Page 106 Nuclear Radiation

1. Three correctly drawn lines **[2]** (1 mark for one correct line)
   Alpha – Highest likelihood of being absorbed and causing damage when passing through living cells.
   Beta – Moderately ionising – can pass through the skin and damage organs inside the body.
   Gamma – Weakly ionising – likely to pass straight through living cells without being absorbed.

2. a) Contamination **[1]**
   b) The sheep had eaten grass that had been contaminated by radioactive particles **[1]**; this made the meat radioactive, so eating it could cause cancer **[1]**

3. Arrows drawn showing alpha being stopped by paper **[1]**; beta being stopped by aluminium **[1]**; and gamma being stopped / partially stopped by lead **[1]**

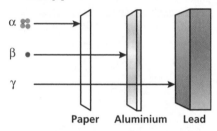

Paper   Aluminium   Lead

Page 107 Using Radioactive Sources

1. a) The count has fallen to a quarter of the original, representing 2 half-lives **[1]**; one half-life = $\frac{4}{2}$ = 2 minutes **[1]**

   b) Background radiation **[1]**
   c) It has a very long half-life of hundreds of years or longer **[1]**

2. a) Medical tracers / radiotherapy **[1]**
   b) Industrial tracers / smoke alarms / thickness control **[1]**

Page 107 Fission and Fusion

1. a) A large, unstable nucleus splits into two or more smaller ones **[1]**; and energy is released **[1]**
   b) Kinetic **[1]**
   c) Neutrons released during fission are absorbed by another nucleus **[1]**; triggering fission in that nucleus, and so on **[1]**
   d) In a nuclear reactor, the reaction is controlled so the amount of nuclei undergoing fission per second is constant **[1]**; in a nuclear explosion, the chain reaction is uncontrolled and each fission triggers more and more events **[1]**

2. Nuclear fission is the splitting of large, unstable nuclei **[1]**; into smaller ones **[1]**; nuclear fusion is the joining together of small nuclei **[1]**; into larger ones **[1]**

Page 108 Stars and the Solar System

1. The expansive force caused by fusion energy/radiation energy **[1]**; and the compression force of gravity **[1]**

2. As it runs out of fuel **[1]**

3. Our sun is relatively small **[1]**; so it is not hot or dense enough to fuse heavier elements **[1]**

4. Iron **[1]**

5. A, C, D, B, E **[4]** (3 marks for three correct; 2 marks for two; 1 mark for one)

6. a) Hydrogen and helium **[1]**
   b) Copper and zinc **[1]**
   c) A blue supergiant is much hotter **[1]**; and denser than the sun **[1]**

Page 109 Orbital Motion and Red-Shift

1. An artificial satellite is manmade, e.g. a TV satellite **[1]**; a natural satellite occurs naturally, e.g. the moon **[1]**

2. Gravity **[1]**

3. a) True **[1]**
   b) True **[1]**
   c) True **[1]**
   d) False **[1]**
   e) False **[1]**

4. a) The Doppler effect **[1]**
   b) As a car moves away, the sound waves are stretched out **[1]**; so have a longer wavelength **[1]**; which means they have a lower frequency **[1]**; and, therefore, a lower pitch **[1]**
   c) Red-shift **[1]**
   d) It will be blue-shifted **[1]**
   e) They are all moving away **[1]**; from Earth **[1]**;
   f) Greater red-shift at greater distances **[1]**; shows that all the galaxies are moving away from one another **[1]**; therefore the universe must be getting bigger **[1]**
   g) The Big Bang theory **[1]**

Pages 110–115 Review Questions

Page 110 Particle Model of Matter

1. Diagrams showing particles in a rigid pattern with no spaces for a solid **[1]**; particles in a random pattern with small spaces, but with particles touching other particles, for a liquid **[1]**; and particles in a random pattern with big spaces between them for a gas **[1]**

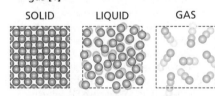

SOLID   LIQUID   GAS

2. a) As a solid is heated, the particles gain energy and begin to vibrate more and more **[1]**; until they are able to move around in a random pattern **[1]**; but not move away from one another **[1]**

b) As a liquid is heated, the particles move faster and faster **[1]**; and eventually have enough energy to move away from each other **[1]**; and escape as a gas **[1]**

3. Measure the mass of the object **[1]**; fill a measuring cylinder to a known mark **[1]**; drop the object into the cylinder and measure the increase in volume **[1]**; divide mass by volume to find density **[1]**

4. a) Initially, the temperature increases **[1]**; then the temperature stays at 0 degrees as energy is being used to melt the ice **[1]**; once all the ice has melted, the temperature increases once more **[1]**; at 100 degrees, the temperature levels off again **[1]**; energy is now being used to turn the water into steam **[1]**; once all of the water has turned into steam, the temperature begins to rise again as more energy is given to the system **[1]**
   b) The pressure will increase **[1]**
   c) $334\,000 \times 0.5$ **[1]**; = 167 000J **[1]**

Page 111 Atoms and Isotopes

1. The electron **[1]**

2. a) Diagram marked with lines showing paths of alpha particles, with some passing straight through **[1]**; some changing direction as they strike the nucleus **[1]**; and some bouncing back **[1]**

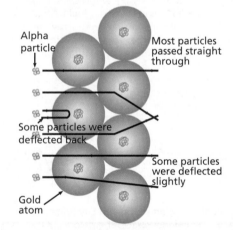

Alpha particle

Most particles passed straight through

Some particles were deflected back

Some particles were deflected slightly

Gold atom

b) Most of the mass of the atoms is concentrated in one place **[1]**; and has a positive charge **[1]**

c) Electrons orbit the nucleus **[1]**; in energy levels / shells **[1]**

3. a) 92 **[1]**
   b) U-233 **[1]**
   c) They both have the same number (92) of protons **[1]**; but U-238 has three more neutrons than U-235 (146 versus 143) **[1]**
   d) An uncharged atom has 92 electrons **[1]**; 92 − 3 = 89 electrons **[1]**

An ion with a charge of +3 has lost three electrons.

Answers

Page 112 Nuclear Radiation

1. Three correctly drawn lines **[2]** (1 mark for one correct line)
   Alpha – A helium nucleus (2 protons and 2 neutrons).
   Beta – An electron emitted from the nucleus.
   Gamma – A high frequency electromagnetic radiation.

2. Because radioactive decay is a random process **[1]**; there is only a probability that a particular nucleus will decay in a certain time **[1]**

Page 112 Using Radioactive Sources

1. Isotope C **[1]**; gamma is the least ionising and can escape the body / alpha is the most ionising and will not be absorbed by the body **[1]**; a half-life of 3 hours is long enough to carry out tests, but not so long that it will remain in the patient for days **[1]**

2. a) The count rate has fallen by a factor of 4 **[1]**; representing 2 half-lives **[1]**; therefore, age = 2 × 6 000 = 12 000 years **[1]**
   b) 18 000 years is equivalent to 3 half-lives **[1]**; so the count will be 8 times lower **[1]**; $\frac{200}{8}$ = 25 counts per minute **[1]**
   c) 60 000 years is ten half-lives **[1]**; and after that time the carbon-14 has completely decayed / its count-rate is no more than background radiation **[1]**

Page 113 Fission and Fusion

1. a) Mass number **[1]**
   b) Alpha decay **[1]**
   c) Beta decay **[1]**
   d) 234 **[1]**; 2 **[1]**
   e) 234 **[1]**; –1 **[1]**

Page 114 Stars and the Solar System

1. Diagram part 1: the gravity pulls together dust and gas **[1]**; and the temperature begins to increase **[1]**; Diagram part 2: the temperature and pressure increase enough that a protostar is formed **[1]**; and it begins to emit light **[1]**; Diagram part 3: the temperature and pressure increase enough that sustained fusion occurs forming a main sequence star **[1]**; the remaining dust forms planets **[1]**

2. Six correctly drawn arrows **[6]**
   Stars the size of the sun – Expands to become a red giant – Fusion stops and the star becomes a white dwarf and eventually a black dwarf
   Larger stars – Expands to become a red giant or red super giant – Explodes in a supernova and then contracts to become a neutron star
   Largest stars – Expands to become a red super giant – Explodes in a supernova and then contracts to become a black hole

3. a) A collection of dust and gas **[1]**
   b) The stage just before the main sequence, when the star is not quite hot enough to begin fusion **[1]**
   c) An explosion that occurs at the end of a star's lifecycle **[1]**

4. The forces of expansion caused by fusion **[1]**; and the collapse due to gravity **[1]**; are in balance **[1]**

5. a) Helium **[1]**
   b) Hydrogen **[1]**

6. Iron **[1]**

Page 115 Orbital Motion and Red-Shift

1. Diagram with arrows correctly drawn to show the direction of force caused by string tension **[1]**; and the direction of instantaneous velocity **[1]**

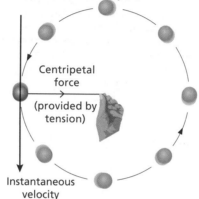

Centripetal force
(provided by tension)

Instantaneous velocity

2. The apparent increase in wavelength of light **[1]**; caused by the object moving away from the observer **[1]**

3. a) The further away the galaxy, the faster it is moving **[1]**
   b) Making measurements over such huge distances is difficult **[1]**; and in the 1920s telescopes were not as good **[1]**
   c) The universe was created in a giant explosion **[1]**; and has been expanding ever since **[1]**
   d) **Any three of:** There are other explanations to explain it **[1]**; it is difficult to take measurements to prove it **[1]**; it is hard to understand how something can explode when there is nothing there **[1]**; their religion tells them something different **[1]**

4. The rate of expansion is increasing **[1]**

Pages 116–127 Mixed Exam-Style Questions

1. a) speed = $\frac{\text{distance}}{\text{time}}$ / $v = \frac{s}{t}$ **[1]**; $= \frac{140}{70}$ **[1]**; = 2m/s **[1]**
   b) 2.5 times as fast (Accept twice as fast) **[1]**
   c) He is stationary **[1]**

2. a) Terminal velocity **[1]**
   b) acceleration = $\frac{\text{change in velocity}}{\text{time taken}}$ / $a = \frac{\Delta v}{t}$ **[1]**; $a = \frac{(10 - 50)}{5}$ **[1]**; = –8m/s² a deceleration of 8m/s² **[1]**
   c) i) As they freefall, they accelerate and the drag force increases **[1]**; eventually the resistive force equals the skydiver's weight **[1]**; so the resultant force is zero and they no longer accelerate **[1]**
      ii) It is harder to move through air when the parachute is open / less streamlined **[1]**; so drag balances / equals weight at a lower speed **[1]**
   d) weight = mass × gravitational field strength / $w = mg$ **[1]**; 75 × 10 = 750N **[1]**

3. a) 200 × 1.5 **[1]**; = 300Nm **[1]**
   b) $\frac{300}{3}$ **[1]**; = 100Nm **[1]**

4. a) Decelerating **[1]**
   b) acceleration = $\frac{\text{change in velocity}}{\text{time taken}}$ / $a = \frac{\Delta v}{t}$ **[1]**; $a = \frac{(50 - 10)}{(6 - 2)}$ **[1]**; $= \frac{40}{4}$ **[1]**; = 10m/s² **[1]**
   c) Distance = area under graph for this region **[1]**; = 50 × 3 **[1]**; = 150m **[1]**

5. speed **[1]**; velocity **[1]**; direction **[1]**

6. a) Towards the centre of the earth **[1]**
   b) Gravity **[1]**
   c) It makes its direction change **[1]**; but its magnitude stays the same **[1]**
   d) The polar orbit **[1]**; because it is closer to the Earth **[1]**
   e) Its mass **[1]**; and speed stay the same **[1]**

7. a) The faster the car, the greater the thinking distance **[1]**; they are directly proportional **[1]**
   b) i) distance = speed × time = 9 × 1 = 9m **[1]**
      ii) 6m **[1]**
      iii) It has increased **[1]**

8. a) work done = force × distance (along the line of action of the force) / $W = Fs$ **[1]**
   b) 400 × 5 **[1]**; = 2000J **[1]**
   c) Kinetic **[1]**
   d) Chemical **[1]**

9. a) i) 15J **[1]**
      ii) 60J **[1]**
      iii) 25J **[1]**
   b) 25% **[1]**
   c) 75% **[1]**

10. a) work done = force × distance / $W = Fs$ **[1]**; W = 250 × 5 = 1250J **[1]**
    b) It is transferred into the kinetic energy of the bicycle **[1]**

11. a) $\frac{3}{12}$ **[1]**; = 0.25m/s² **[1]**
    b) power = $\frac{\text{energy transferred}}{\text{time}}$ / $P = \frac{W}{t}$ **[1]**; $P = \frac{2400}{12}$ = 200W **[1]**

Answers

c) resultant force = mass × acceleration / $F = ma$ [1]; force used on first car = 800 × 0.25 = 200N [1]; acceleration of second car = $\frac{3}{18}$ = 0.17m/s² [1]; mass of second car = $\frac{200}{0.17}$ = 1200kg [1]

12. a) Nuclear provided slightly more energy than renewables (5.5% more of the national share) [1]
b) **Any two of:** wind turbines [1]; solar panels [1]; waves [1]; tides [1]; geothermal [1]
c) Advantage: reliable / high output / no greenhouse gases produced / large available fuel supply [1]; disadvantage: radioactive waste / expensive to build and decommission / risk of nuclear accident [1]
d) gravitational energy [1]; → kinetic energy [1]
e) It reduces energy loss on transmission, as the cables don't get hot [1]; but higher voltages are more dangerous and must be transformed back to safer voltages at the other end [1]

13. a) 1900 × 4.5 [1]; = 8550J [1]
b) $\frac{8500}{9}$ [1]; = 950W [1]
c) 950 × 2 [1]; = 1900J [1]

14. When the current becomes too big [1]; the electromagnet becomes powerful enough that it exerts a force bigger than the spring [1]; pushing the bolt to the right [1]; this attracts the iron bolt [1]; with the iron bolt and plunger removed, the spring pushing upwards lifts the switch breaking the circuit [1]

15. a) i) light [1]; electrical [1]
   ii) electrical [1]; chemical [1]
   iii) electrical [1]; kinetic [1]
b) i) $\frac{1500}{10}$ [1]; = 150W [1]
   ii) Heat / sound [1]

16. a) It is reduced [1]
b) To reduce energy loss on transmission [1]
c) Step-down transformer [1]

17. a) Iron [1]
b) Ratio of turns = $\frac{46\,000}{4000}$ = 11.5 [1];

output voltage = $\frac{230}{11.5}$ [1]; = 20V [1]
(Accept calculations using the full transformer equation, which result in the correct answer)

The voltage on the secondary coil is 11.5 times smaller than the voltage on the primary coil.

18. a) 800W [1]
b) $\frac{800}{230}$ [1]; = 3.48A [1]
c) 5A [1]

19. a) Correctly labelled voltmeter [1]; diode [1]; and ammeter [1]

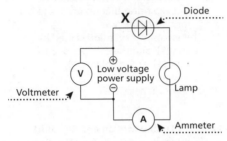

b) When the voltage is negative, the resistance is extremely high [1]; and no current flows [1]; with a positive voltage, the resistance is extremely high at first, so no current flows [1]; but after a certain point, the resistance falls and current flows [1]

20. a) Circuit with correctly connected battery [1]; ammeter [1]; and voltmeter [1]

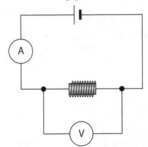

b) i) resistance = $\frac{\text{potential difference}}{\text{current}}$ / $R = \frac{V}{I}$ [1]; $R = \frac{9}{0.30}$ = 30Ω [1]
   ii) The resistance has increased [1]

21. a) potential difference = current × resistance / $V = IR$ [1]; 0.5 × 8 = 4V [1]
b) 9 – 4 = 5V [1]
c) 0.5A [1]

22. a) A1 = 0.5A [1]; A4 = 0.5A [1]
b) 20 × 0.3 [1]; = 6V [1]
c) 6V [1]

23. a) Microwaves are absorbed by food [1]; when this happens energy is transferred from the microwaves to the food [1]; causing its temperature to increase [1]
b) wave speed = frequency × wavelength / $v = f\lambda$ [1]; v = 245 000 000 × 0.122 [1]; = 29 890 000m/s or 2.989 × 10⁷ [1]

24. a) Correctly drawn continuation of ray AB after entering the block [1]; and CD after entering the block [1]
b) Correctly drawn lines indication wave fronts [1]; that are closer together than those outside the block [1]

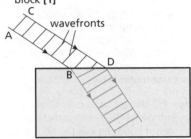

c) The wave slows down when it enters the block, reducing the wavelength [1]; the part of the wave that enters first, slows down first, causing the wave to bend towards the normal [1]

25. a) The normal [1]
b) This prevents the ray from being refracted on entry to the block [1]; so that accurate measurements can be made [1]
c) If they were directly proportional, doubling one would double the other [1]; when A is doubled from 30° to 60°, B goes from 19° to 35°, so it is not doubled [1] (accept any other example taken from data in table)
d) There is a positive correlation between Angle A and Angle B [1]; as A increases, so does B [1]
e) These are the limits of the data / the only measurements taken [1]; what happens on either side of the data range is unknown [1]

26. a) Weight [1]
b) It accelerates [1]
c) Its acceleration [1]; increases [1]
d) i) Time taken to reach 8100m/s from stationary = 9 × 60 = 540s [1];

acceleration = $\frac{\text{change in velocity}}{\text{time taken}}$ / $a = \frac{\Delta v}{t}$ [1]; $a = \frac{8100}{540}$ = 15m/s² [1]

ii) The velocity is the speed in a specific direction [1]; with speed, direction does not matter [1]

Notes

Physics Equations

You must be able to recall and apply the following equations using standard units:

Word Equation	Symbol Equation
weight = mass × gravitational field strength	$W = mg$
work done = force × distance (along the line of action of the force)	$W = Fs$
force applied to a spring = spring constant × extension	$F = ke$
moment of a force = force × distance (normal to direction of force)	$M = Fd$
pressure = $\dfrac{\text{force normal to a surface}}{\text{area of that surface}}$	$p = \dfrac{F}{A}$
distance travelled = speed × time	$s = vt$
acceleration = $\dfrac{\text{change in velocity}}{\text{time taken}}$	$a = \dfrac{\Delta v}{t}$
resultant force = mass × acceleration	$F = ma$
HT momentum = mass × velocity	$p = mv$
kinetic energy = 0.5 × mass × (speed)²	$E_k = \tfrac{1}{2}mv^2$
gravitational potential energy = mass × gravitational field strength × height	$E_p = mgh$
power = $\dfrac{\text{energy transferred}}{\text{time}}$	$P = \dfrac{E}{t}$
power = $\dfrac{\text{work done}}{\text{time}}$	$P = \dfrac{W}{t}$
efficiency = $\dfrac{\text{useful output energy transfer}}{\text{total input energy transfer}}$	
efficiency = $\dfrac{\text{useful power output}}{\text{total power input}}$	
wave speed = frequency × wavelength	$v = f\lambda$
charge flow = current × time	$Q = It$
potential difference = current × resistance	$V = IR$
power = potential difference × current	$P = VI$
power = (current)² × resistance	$P = I^2R$

Word Equation	Symbol Equation
energy transferred = power × time	$E = Pt$
energy transferred = charge flow × potential difference	$E = QV$
density = $\dfrac{\text{mass}}{\text{volume}}$	$\rho = \dfrac{m}{V}$

The following equations will appear on the equations sheet that you are given in the exam. You must be able to select and apply the appropriate equation to answer a question correctly.

Word Equation	Symbol Equation
HT pressure due to a column of liquid = height of column × density of liquid × gravitational field strength	$p = h\rho g$
(final velocity)² – (initial velocity)² = 2 × acceleration × distance	$v^2 - u^2 = 2as$
HT force = $\dfrac{\text{change in momentum}}{\text{time taken}}$	$F = \dfrac{m\Delta v}{\Delta t}$
elastic potential energy = 0.5 × spring constant × (extension)²	$E_e = \frac{1}{2}ke^2$
change in thermal energy = mass × specific heat capacity × temperature change	$\Delta E = mc\Delta\theta$
magnification = $\dfrac{\text{image height}}{\text{object height}}$	
period = $\dfrac{1}{\text{frequency}}$	
HT force on a conductor (at right-angles to a magnetic field) carrying a current = magnetic flux density × current × length	$F = BIl$
thermal energy for a change of state = mass × specific latent heat	$E = mL$
HT $\dfrac{\text{potential difference across primary coil}}{\text{potential difference across secondary coil}} = \dfrac{\text{number of turns in primary coil}}{\text{number of turns in secondary coil}}$	$\dfrac{V_p}{V_s} = \dfrac{n_p}{n_s}$
HT potential difference across primary coil × current in primary coil = potential difference across secondary coil × current in secondary coil	$V_sI_s = V_pI_p$
For gases: pressure × volume = constant	$pV = constant$

Glossary and Index

Focal length the distance from the centre point of a lens to the focus point, where the light rays converge / come together   **38–39**

Force an influence that occurs when two objects interact   **8–9**

HT Free body diagram a diagram used to show the relative magnitude and direction of all the forces acting on an object in a given situation   **9**

Frequency the number of times that a wave / vibration repeats itself in a specified time period   **30**

Fusion a reaction in which two nuclei combine to form a nucleus with the release of energy   **93**

G

Galaxy a collection of millions / billions of stars held together by gravitational attraction   **94**

Gamma high frequency, short wavelength electromagnetic waves; a type of nuclear radiation, emitted from a nucleus   **36–37, 88–89**

Gear a wheel with teeth that engages with another wheel with teeth, or with a rack, in order to change the speed or direction of transmitted motion   **11**

HT Generator effect to induce a potential difference (voltage) using magnetic fields and conductors   **70–71**

Gradient a measure of the steepness of a sloping line; the ratio of the change in vertical distance over the change in horizontal distance   **17**

Gravitational potential energy (GPE) the energy gained by raising an object above ground level (due to the force of gravity)   **26**

Gravity the force of attraction exerted by all masses on other masses, only noticeable with a large body, e.g. the Earth or Moon   **9**

H

Half-life the average time it takes for half the nuclei in a sample of radioactive isotope to decay; the time it takes for the count rate / activity of a radioactive isotope to fall by **50%** (halve)   **90**

I

Incompressible cannot be compressed / squeezed / compacted   **84**

Induce to produce (a potential difference) or transmit (magnetism)   **66–67, 70–71**

Induced magnet an object that becomes magnetic when placed in a magnetic field   **66–67**

Inelastically deformed describes an object that cannot return to its original shape when the forces that caused it to change shape are removed (because the limit of proportionality has been exceeded)   **10**

HT Inertia the tendency of a body to stay at rest or in uniform motion unless acted upon by an external force   **15, 17**

Infrared the part of the electromagnetic spectrum with a longer wavelength than light but a shorter wavelength than radio waves   **36–37**

Internal energy the sum of the energy of all the particles that make up a system, i.e. the total kinetic and potential energy of all the particles added together   **26–27**

Inversely proportional a relationship between two variables, where one variable increases and the other decreases   **17**

Ion formed when an atom loses or gains one or more electrons to become charged   **86**

Ionising refers to radiation that can cause atoms to lose or gain atoms, becoming ions   **37**

Irradiation to expose an object to nuclear radiation (the object does not become radioactive)   **89**

Isolated refers to an object that has no conducting path to earth   **64–65**

Isotopes atoms of the same element, but with different numbers of neutrons in the nucleus   **86**

K

Kinetic energy the energy of motion of an object, equal to the work it would do if brought to rest   **26**

L

Latent heat of fusion the amount of heat energy needed for a specific amount of substance to change from solid to liquid   **85**

Latent heat of vaporisation the amount of heat energy needed for a specific amount of substance to change from liquid to gas   **85**

Lever a rigid bar set on a pivot, used to transfer a force to a load   **11**

Limit of proportionality the point up to which the extension of an elastic object is directly proportional to the applied force (once exceeded the relationship is no longer linear)   **10**

Longitudinal a wave in which the oscillations are parallel to the direction of energy transfer, e.g. sound waves   **30**

M

Magnification the ratio of image height to object height, e.g. $\text{magnification} = \dfrac{\text{image height}}{\text{object height}}$ **38**

Main sequence a stable period in the lifecycle of a star, during which the outward acting fusion energy and inward acting force of gravity are in balance   **94**

Mass a measure of how much matter an object contains, measured in kilograms (kg)   **9**

Medium a material or substance   **32**

Meniscus the curved upper surface of a liquid standing in a tube, from which measurements of volume are taken   **84**

HT Microphone a device that uses the generator effect to convert soundwaves into electrical signals   **71**

Microwaves electromagnetic radiation in the wavelength range 0.3 to 0.001 metres, used in satellite communication and cooking   **36–37**

Moment a measure of the turning effect of a force that causes an object to rotate about a pivot point, calculated by multiplying force by distance   **11**

HT Momentum the product of an object's mass and velocity   **18**

HT Motor effect the force experienced by a current carrying conductor when it is placed in a magnetic field, which is used to create movement in electrical motor   **68–69**

N

National Grid the network of high voltage power lines and transformers that connects major power stations, businesses and homes   **63**

Nebula a cloud of particles and gases   **94–95**

Neutron a neutral subatomic particle; a type of nuclear radiation, which can be emitted during radioactive decay   **86–87**

Non-contact force a force that occurs between two objects that are not in contact (not touching), e.g. gravitational and electrostatic forces   **8**

Normal at right-angles to / perpendicular to   **12–13, 33**

Nucleus the positively charged, dense region at the centre of an atom, made up of protons and neutrons, orbited by electrons   **86–87**

O

Ohmic conductor a resistor in which the current is directly proportional to the potential difference at a constant temperature   **57**

Opaque describes an object that either reflects or absorbs all light incident on its surface, so that no light passes through it   **40**

Orbit the curved path followed by a planet, satellite, comet, etc. as it travels around a body that exerts a gravitational force upon it; also applies to the paths of electrons around a nucleus (which exerts an electrostatic force of attraction)   **96**

Oscillate to vibrate / swing from side to side with a regular frequency   **30–31**

P

Parallel (circuit) a circuit in which the components are connected side by side on a separate branch / path, so that the current from the cell / battery splits with a portion going through each component   **58**

Particle an extremely small body with finite mass and negligible (insignificant) size, e.g. protons, neutrons and electrons   **84**

Period   the time taken for a wave to complete one oscillation; the time it takes for a particle in the wave to move backwards and forwards once around its undisturbed position   30

Permanent magnet   an object that produces its own magnetic field   66

Pivot   the point around which an object turns   11

Pole   the two opposite regions in a magnet, where the magnetic field is concentrated; can be north or south   66–67

Potential difference   the difference in electric potential between two points in an electric field; the work that has to be done in transferring a unit of positive charge from one point to another, measured in volts (V)   55

Power   a measure of the rate at which energy is transferred or work is done   59, 61

Precision   refers to the degree of accuracy (of a measuring instrument), i.e. the minimum possible change that can be measured   55

Pressure   the force exerted on a surface, e.g. by a gas on the walls of a container   12–13

Principal axis   the horizontal line that runs straight through the centre of a lens   38–39

Principal focus   (also called *focal point*) the point where parallel rays of light travelling through a lens converge (meet) or from which they appear to diverge (spread out) from refraction by the lens   38–39

Proportional   describes two variables that are related by a constant ratio   17

Proton   a subatomic particle found in the nucleus of an atom, with an electrical charge of +1   86–87

Protostar   the hot dense mass that forms when a nebula collapses, which evolves into a star once nuclear fusion can occur   95

P-waves   (*Primary waves*) the longitudinal seismic waves produced during an earthquake   35

R

Radioactive   containing a substance that gives out radiation   88–89

Real   refers to an image produced by a lens, which is on the opposite side of the lens to the object and can be projected onto a screen (opposed to a *virtual image*)   38–39

Red-shift   the observed increase in the wavelength of light from distant galaxies, towards the red end of the spectrum   97

Reflected   when a wave meets a boundary between two different materials and is bounced back   32

Refracted   when a wave meets a boundary between two different materials and changes direction   32–33

Refractive index   a measure of the extent to which light is refracted by a material   32

Renewable   can be replaced   29

Reproducible   results are reproducible if the investigation / experiment can be repeated by another person, or by using different equipment / techniques, and the same results are obtained, demonstrating that the results are reliable   27

Repulsion   a force that pushes two objects apart, such as the force between two like electric charges or magnetic poles   64–65

Resistance   a measure of how a component resists (opposes) the flow of electrical charge, measured in ohms (Ω)   55–57

Resultant   a single force that represents the overall effect of all the forces acting on an object   9

S

Scalar   a quantity, such as time or temperature, that has magnitude but no direction   8

Seismic   caused by an earthquake   35

Series (circuit)   describes a circuit in which the components are connected one after then other, so the same current flows through each component   59

Solenoid   a coil of wire partially surrounding an iron core, which is made to move inside the coil by the magnetic field set up by a current; used to convert electrical to mechanical energy   67

Spark   a momentary flash of light accompanied by a sharp crackling noise, produced by a sudden electrical discharge through the air between two points   64–65

Specific heat capacity   the amount of energy required to raise the temperature of one kilogram of substance by one degree Celsius   26–27

Specular reflection   reflection in a single direction (no scattering of light)   40

Speed   a scalar measure of the distance travelled by an object in a unit of time, measured in metres per second (m/s)   14–15

Spring constant   a measure of how easy it is to stretch or compress a spring; calculated as: $\dfrac{\text{force}}{\text{extension}}$ 10–11

State of matter   the structure and form of a substance, i.e. gas, liquid or solid   84

Supernova   a large star that explodes because the forces within it are no longer balanced / stable, releasing vast amounts of energy   95

S-waves   (*Secondary waves*) the transverse seismic waves produced during an earthquake   35

System   an object or group of objects   26–27

T

Terminal velocity   the constant maximum velocity reached by a body falling under gravity   18

Tracer   a radioactive isotope that is put into a system so that its movement can be tracked, helping to reveal blockages / holes that should not be there; used in medicine and industry   91

Transferred   refers to how energy is changed, e.g. chemical energy can be transferred to electric energy   28–29

Transformer   a device that transfers an alternating current from one circuit to another, with an increase (step-up transformer) or decrease (step-down transformer) of voltage   63, 71

Translucent   an object that transmits light, but scatters the rays so that objects cannot be seen clearly through it   40

Transmitted   when waves are sent out from a source or pass through a material   32

Transparent   an object that transmits light coherently (the light rays do not get scattered), so that objects on the other side can be seen clearly   40

Transverse   a wave in which the oscillations are at right-angles to the direction of energy transfer, e.g. water waves   30

U

Ultrasonic   sound waves with a frequency greater than 20kHz, so they cannot be heard by humans   34

Ultraviolet   the part of the electromagnetic spectrum with wavelengths shorter than light but longer than X-rays   36–37

Uniform   unchanging in form, quality or quantity; regular   67

Universe   all existing matter, energy and space   94

Unstable   lacking stability; having a very short lifetime; radioactive   88

HT **Upthrust**   an upward push; the upwards force exerted by a fluid on an object in / partially in the fluid   13

V

Vector   a variable quantity that has magnitude and direction   8–9

Velocity   a vector quantity that provides a measure for the speed of an object in a given direction   15

Virtual   refers to an image produced by a lens, which is on the same side of the lens as the object and can only be seen by looking through the lens (opposed to a *real image*)   38–39

W

Wavelength   the distance from one point on a wave to the equivalent point on the next wave, measured in metres (m), represented by the symbol λ 30–31

Weight   the vertical downwards force acting on an object due to gravity   9

Work   the product of force and distance moved along the line of action of the force, when a force causes an object to move   10

X

X-rays   the part of the electromagnetic spectrum with wavelengths shorter than that of ultraviolet radiation but longer than gamma rays   36–37

Collins

AQA GCSE 9-1

Physics

Workbook

Nathan Goodman

Physics

AQA GCSE 9-1 Workbook

Revision Tips

Rethink Revision

Have you ever taken part in a quiz and thought *'I know this!'* but, despite frantically racking your brain, you just couldn't come up with the answer?

It's very frustrating when this happens but, in a fun situation, it doesn't really matter. However, in your GCSE exams, it will be essential that you can recall the relevant information quickly when you need to.

Most students think that revision is about making sure you **know** stuff. Of course, this is important, but it is also about becoming confident that you can **retain** that *stuff* over time and **recall** it quickly when needed.

Revision That Really Works

Experts have discovered that there are two techniques that help with all of these things and consistently produce better results in exams compared to other revision techniques.

Applying these techniques to your GCSE revision will ensure you get better results in your exams and will have all the relevant knowledge at your fingertips when you start studying for further qualifications, like AS and A Levels, or begin work.

It really isn't rocket science either – you simply need to:

- **test yourself** on each topic as many times as possible
- **leave a gap** between the test sessions.

Three Essential Revision Tips

1. **Use Your Time Wisely**

   - Allow yourself plenty of time.
   - Try to start revising at least six months before your exams – it's more effective and less stressful.
   - Your revision time is precious so use it wisely – using the techniques described on this page will ensure you revise effectively and efficiently and get the best results.
   - Don't waste time re-reading the same information over and over again – it's time-consuming and not effective!

2. **Make a Plan**

   - Identify all the topics you need to revise (this All-in-One Revision & Practice book will help you).
   - Plan at least five sessions for each topic.
   - One hour should be ample time to test yourself on the key ideas for a topic.
   - Spread out the practice sessions for each topic – the optimum time to leave between each session is about one month but, if this isn't possible, just make the gaps as big as realistically possible.

3. **Test Yourself**

   - Methods for testing yourself include: quizzes, practice questions, flashcards, past papers, explaining a topic to someone else, etc.
   - This All-in-One Revision & Practice book provides seven practice opportunities per topic.
   - Don't worry if you get an answer wrong – provided you check what the correct answer is, you are more likely to get the same or similar questions right in future!

Visit our website to download your free flashcards, for more information about the benefits of these techniques, and for further guidance on how to plan ahead and make them work for you.

www.collins.co.uk/collinsGCSErevision

Contents

Forces – An Introduction

1 When a plane is in flight, the engines provide a thrust force that pushes the aircraft forwards. The wings provide a 'lift' force that acts upwards.

a) Name **two** other forces that act on the plane.

In each case state whether it is contact force or non-contact force.

....

.... [4]

b) The plane has a mass of 120 000kg. Calculate the weight of the plane (g = 10N/kg).

Weight = ... N [2]

c) The plane accelerates and ascends to a higher altitude.
During this time, the resultant forwards force is twice the size of the resultant upwards force.

Use this information to draw a scale
vector diagram.
Your diagram should show:
• The resultant forwards force.
• The resultant upwards force.
• A final resultant force that shows the
  combined effect of all the forces acting
  on the plane. [3]

2 A student carries out an investigation into forces.
They use an air track, which suspends a glider vehicle on a cushion of air so that it can move smoothly without touching the ground.

a) What force is the air track designed to reduce?

Answer .... [1]

b) Use the idea of contact and non-contact forces to explain why the air track is effective at doing this.

....

.... [2]

Total Marks .... / 12

Forces in Action

1 A student carries out an investigation involving springs.

The student suspends a spring from a rod.

A force is applied to the spring by suspending a mass from it.

a) Describe how the student could test if the spring is behaving elastically.

.... [2]

....

b) In a second investigation, the student takes a set of measurements for force and extension.
The results are shown in **Table 1** below.

Table 1

Force (N)	0.0	1.0	2.0	3.0	4.0	5.0	6.0
Extension (cm)	0.0	4.0		12.0	16.0	22.0	31.0

i) Add the missing value to the table. [1]

ii) Explain why you chose this value.

....

.... [2]

2 A man uses a long lever to remove tree stump from the ground.

The total length of the lever is 1.8m and the lever is pivoted 20cm from the end.

Figure 1

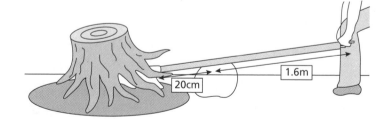

20cm   1.6m

a) The man pushes down with a force of 200N.

Work out the turning moment created.

Answer .... [2]

b) When pressing down with a force of 200N the man is just able to lift the tree stump.

Use this information and your answer to part **a)** to work out the force needed to lift the stump.

Answer .... [2]

Total Marks .... / 9

Pressure and Pressure Differences

1 A hydraulic jack uses pressure to transfer a force and lift a car.
Figure 1 shows the construction of the jack.
The pressure in the liquid is the same at all points.

Figure 1

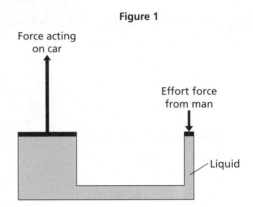

a) Give the equation that links pressure, force and area.

Answer _____ [1]

b) Use **Figure 1** to explain how the man using the jack is able to exert a much greater force on the car than he would be able to without using the jack.

_____

_____

_____ [3]

2 A hot air balloon experiences a force of upthrust caused by the surrounding air.
This force causes the hot air balloon to rise.

a) What can be said about the relative density of the hot air balloon and the air around it?

_____

_____ [1]

b) The hot air balloon rises until it reaches a steady height.

Use your understanding of forces and the atmosphere to explain why this happens.

_____

_____

_____

_____ [3]

Total Marks _____ / 8

Forces and Motion

1 A person takes their dog for a walk.

The graph in **Figure 1** shows how the distance from their home changes with time.

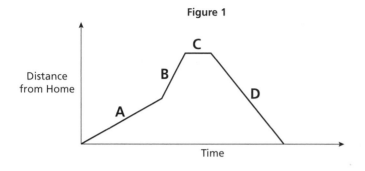

Figure 1

a) Work out the final displacement at the end of their journey.

Answer _____ [1]

b) Which part of the graph, **A**, **B**, **C** or **D**, shows them walking at the fastest rate?

Answer _____ [1]

c) Describe their motion during section **C**.

Answer _____ [1]

d) Describe how the velocity of section **A** compares to the velocity of section **D**.

_____

_____

_____ [3]

2 Drivers on a racetrack enter a hairpin bend travelling east.

The tight bend forces them to slow down.

When they exit the bend, they are travelling west and speed back up again.

The bend is 180m long.

a) If it takes 6 seconds to travel around the bend, what is the average speed of the car around the bend?

Speed = _____ m/s [2]

b) A driver enters the bend at 50m/s and exits the bend at 40m/s.

i) Work out the change in speed.

Speed = _____ m/s [1]

ii) Work out the change in velocity.

Speed = _____ m/s [1]

Total Marks _____ / 10

Forces and Acceleration

1 An experiment is carried out to investigate how changing the mass affects the acceleration of a system.

A trolley is placed on a bench and is made to accelerate by applying a constant force using hanging masses. Different masses were then added to the trolley.

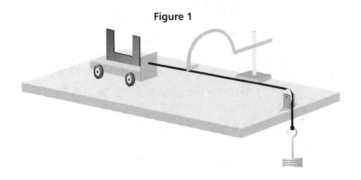

Figure 1

a) What is the independent variable?

   Answer _____ [1]

b) What is the control variable?

   Answer _____ [1]

c) What is the dependent variable?

   Answer _____ [1]

2 A boat accelerates at a constant rate in a straight line.
This causes the velocity to increase from 4.0m/s to 16.0m/s in 8.0s.

a) Calculate the acceleration.
   Give the unit.

   Answer _____ [3]

b) A water skier being pulled by the boat has a mass of 68kg.

   Use your answer to part **a)** to calculate the resultant force acting on the water skier whilst accelerating.

   Answer _____ [2]

3 The manufacturer of a car gives the following information in a brochure:
The mass of the car is 950kg.
The car will accelerate from 0 to 33m/s in 11s.

a) Calculate the acceleration of the car during the 11s.

   Answer _____ [2]

b) Calculate the force needed to produce this acceleration.

   Answer _____ [2]

Total Marks _____ / 12

Terminal Velocity and Momentum

1 The terminal velocity is the maximum velocity a falling object can reach.

Describe how a skydiver could increase their terminal velocity.

....

.... [2]

2 A Saturn 5 rocket, as used on the Apollo space missions, has F1 rocket engines.
Each engine burns 5000kg of fuel every second, producing a thrust of 6.7MN.

a) Given that **force = rate of change of momentum**, calculate the speed with which the exhaust gases exit each engine.

Answer .... m/s [3]

b) During the Apollo mission, five engines were used simultaneously.

Use your answer to part **a)** to calculate the momentum gained by the rocket in the first 10 seconds of travel (assuming there were no other resistive forces).

Answer .... kg m/s [3]

c) The entire launch vehicle has a mass of 3000 000kg.

Use your answer to part **b)** to calculate the velocity after the first 10 seconds (assuming all other factors are unchanged).

Answer .... m/s [3]

d) The rocket engines provide constant thrust.
However, after 100 seconds, the acceleration of the rocket is greater than the initial acceleration.

Give **two** reasons why this might be.
You must explain your answers.

....

....

....

.... [4]

Total Marks .... / 15

Stopping and Braking

1 The graph in **Figure 1** shows how the velocity of a car changes from the moment the driver sees an obstacle blocking the road.

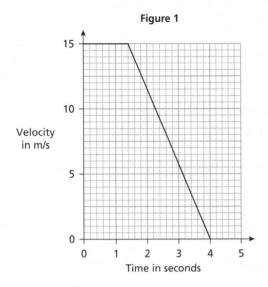

Figure 1

a) Work out the reaction time of the driver.

Answer .... [1]

b) Use the graph and your answer to part **a)** to calculate the thinking distance.

Answer .... [2]

c) Use the graph to work out the time it took for the car to stop from the moment the brakes were applied.

Answer .... [1]

d) The car and driver have a combined mass of 1300kg.

Work out the momentum of the car just before the brakes were applied.

Answer .... [2]

e) Use your answers to parts **c)** and **d)** to work out the braking force applied by the brakes of the car.

Answer .... [3]

f) The driver of the car was tired and had been drinking alcohol.

On the graph in **Figure 1**, sketch a second line to show how the graph would be different if the driver had been wide awake and fully alert. [3]

g) How would the graph look different if the vehicle had old / worn brakes and tyres?

....

.... [2]

Total Marks .... / 14

Energy Stores and Transfers

1 An electric kettle is used to heat 2kg of water from 20°C to 100°C.

> change in thermal energy = mass × specific heat capacity × temperature change
>
> Specific heat capacity of water = 4200J/kg°C

a) All of the energy supplied to the kettle goes into the water.

Calculate the amount of electrical energy supplied to the kettle.

Answer .... [3]

b) On a different occasion, the kettle is filled with 2.5kg of water but is switched on for the same amount of time.

Use your answer to part **a)** to calculate what temperature the kettle heats the water to in this period.

Answer .... [3]

2 **Figure 1** shows a pendulum swinging backwards and forwards.
The pendulum is made from a 100g mass suspended by a light string.

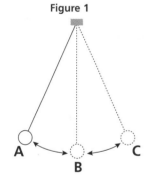

Figure 1

a) At position **C**, the mass is 5cm higher than at position **B**.

Calculate the difference in gravitational energy between positions **B** and **C**.
The gravitational field strength is = 10N/kg.

Answer .... [3]

b) The difference in potential energy is the same as the amount of kinetic energy gained by the mass as it swings down from **C** to **B**.

Calculate the velocity of the mass at position **B**.

Answer .... [3]

Total Marks .... / 12

Energy Transfers and Resources

1 Complete the sentences below to explain the energy transfers involved in a solar panel.

In a solar panel, _____ energy is converted into _____ energy.

Some energy is converted into _____ energy and lost to the surroundings. **[3]**

2 A student tested four different types of fleece, **J**, **K**, **L** and **M**, to find out which would make the warmest jacket.
Each type of fleece was wrapped around a can. The can was then filled with hot water.
The temperature of the water was recorded every 2 minutes for a 20-minute period.
The graph in **Figure 2** shows the student's results.

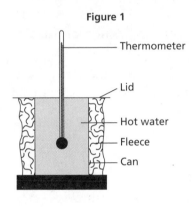

Figure 1

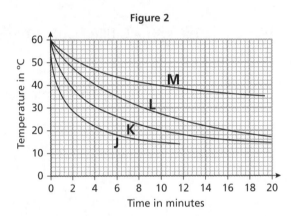

Figure 2

a) To be able to compare the results, it was important to use the same volume of water in each test.

Give **two** other variables that should have been kept the same in each test.

_____ **[2]**

b) Which type of fleece, **J**, **K**, **L** or **M**, should the student recommend for making a ski jacket? You must explain your answer.

_____

_____ **[2]**

Total Marks _____ / 7

Waves and Wave Properties

1 A wave machine in a swimming pool generates waves with a frequency of 0.5Hz.

a) What does a frequency of 0.5Hz mean?

.... [1]

b) Give the equation that links the frequency, speed and wavelength of a wave.

.... [1]

c) The swimming pool is 50m long.
It takes each wave 10 seconds to travel the length of the pool.

Calculate the wave speed.

Answer .... [2]

d) Use your answers to parts **b)** and **c)** to calculate the wavelength of the waves.

Answer .... [2]

e) One section of the swimming pool is designed for young children and has much shallower water.
A parent notices that the waves get closer together when they enter this section.

What effect will this have on the wave speed?

.... [1]

2 Waves may be longitudinal or transverse.

Describe the differences between longitudinal and transverse waves.

....

....

....

....

....

.... [3]

Total Marks .... / 10

Reflection, Refraction and Sound

1   **Figure 1** shows two beakers.
    Each beaker has a drawing pin inside.

    The first beaker is empty and the eye cannot
    see the drawing pin.
    The second beaker is full of water and the
    drawing pin can be seen.

    Explain how this is possible.
    You may add a ray line to the diagram to help with your answer.

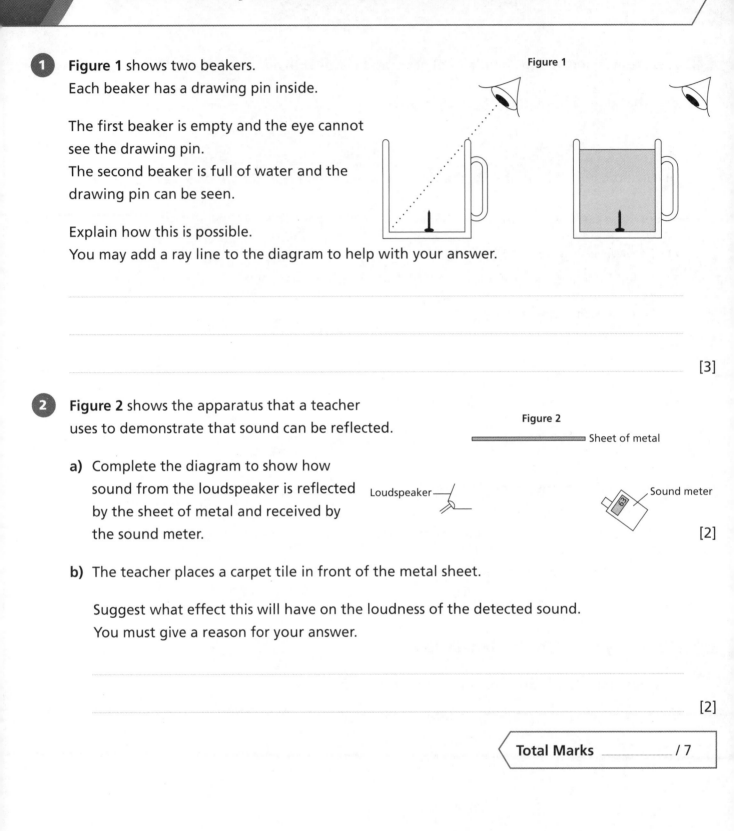

Figure 1

....

....

....

                                                                                        [3]

2   **Figure 2** shows the apparatus that a teacher
    uses to demonstrate that sound can be reflected.

    Figure 2

    a) Complete the diagram to show how
       sound from the loudspeaker is reflected
       by the sheet of metal and received by
       the sound meter.   [2]

    b) The teacher places a carpet tile in front of the metal sheet.

       Suggest what effect this will have on the loudness of the detected sound.
       You must give a reason for your answer.

       ....

       ....

                                                                                        [2]

    Total Marks _____ / 7

Waves for Detection and Exploration

1 Ultrasound can be used to measure the depth of water below a ship.
A pulse of ultrasound is emitted by an electronic system on-board the ship.
It takes 0.8 seconds for the ultrasound to be received back at the ship.

Calculate the depth of the water.
Speed of ultrasound in water = 1600m/s

Answer .... [3]

2 Ultrasound can also be used to detect internal cracks in materials.
An ultrasound pulse is transmitted into the material and any reflected pulses are measured by a detector.
Figure 1 shows the screen of an oscilloscope connected to a detector, which is being used to examine a steel block.

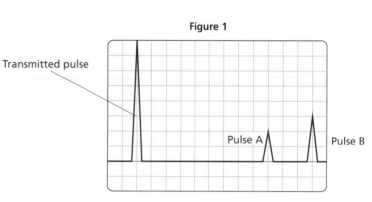

Figure 1

a) Pulse **A** is reflected back by an internal crack.

What is Pulse **B** reflected back from?

Answer .... [1]

b) The metal block is 120mm from front to back.

Study the oscilloscope trace and work out the distance, in millimetres, from the front of the block to the internal crack.

Answer .... [1]

c) When investigating a second steel block, the oscilloscope settings are left exactly the same. Pulse A is located at the same position, but the amplitude of the reflected pulse is 1 square high.

What conclusion can be drawn about the size of the crack in the second steel block?

.... [1]

Total Marks .... / 6

The Electromagnetic Spectrum

1   Radio waves and visible light are electromagnetic waves that are used for communication.

   a) Name another type of electromagnetic wave that is used for communication.

   Answer _____ [1]

   b) Name an electromagnetic wave which is **not** used for communication and give one of its uses.

   _____

   _____ [2]

2   Different parts of the electromagnetic spectrum have different uses.
   Figure 1 shows the electromagnetic spectrum.

Figure 1

Radio waves	Microwaves	Infrared	Visible light	Ultraviolet	X-rays	Gamma rays

Complete the sentence below using words from the box.

amplitude   frequency   speed   wavelength

The arrow in the diagram points in the direction of increasing _____

and decreasing _____ .   [2]

3   After a person is injured, a doctor will sometimes request a photograph of the patient's bones.

   a) Which type of electromagnetic radiation would be used to produce the photograph?

   Answer _____ [1]

   b) What properties of this radiation enable it to be used to photograph bones?

   _____

   _____

   _____ [2]

Total Marks _____ / 8

Lenses

1 The magnification of a lens can be found by dividing the actual height of object by the height of the image.

   a) A lens produces a 2cm tall image of an 8cm tall object.

   What is the magnification of the lens?   Answer .... [2]

   b) A cinema projector produces a 3m tall image from a 2cm tall object.

   What is the magnification of the projector lens?   Answer .... [2]

2 Explain why concave lenses are also called diverging lenses.

   ....

   .... [2]

3 A lens is used to produce an image.
   The image produced is real, inverted and magnified.

   a) What kind of lens has been used?   Answer .... [1]

   b) Where is the image in relation to the lens?
      Tick **one** box.

      Closer to the lens than the object   ☐

      More distant from the lens than the object   ☐

      The same distance from the lens as the object   ☐   [1]

4 **Figure 1** shows the position of an object formed by a lens.

   a) What type of lens is shown in **Figure 1**?

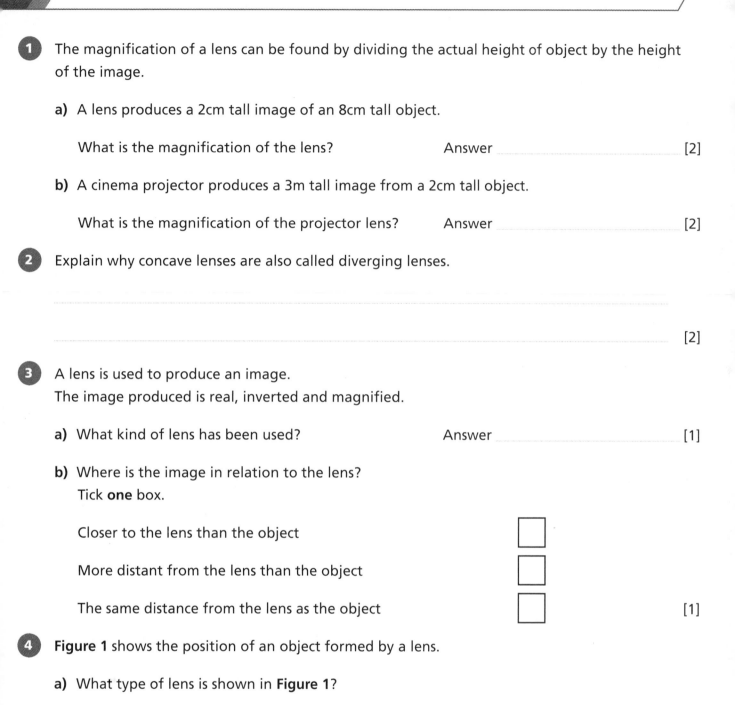

Figure 1

   Answer .... [1]

   b) Construct a ray diagram on **Figure 1** to show how the image is formed. [4]

   c) Use your ray diagram to calculate the magnification of the lens.

   Magnification = ... [2]

Total Marks .... / 15

Light and Black Body Radiation

1 A trainee lighting technician is trying to mix red and green light.
The trainee places a red and a green filter in front of a white spotlight.
Figure 1 shows how this has been set up.

Figure 1

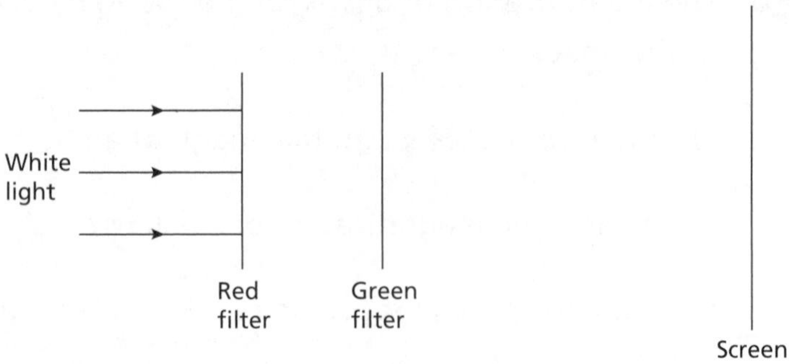

a) No light reaches the screen.

Explain why.

....

....

....

[2]

b) The trainee lighting technician cuts a circle in the red filter.

What will be seen on the screen now?

.... [1]

2 An astronomer is studying distant stars using an infrared telescope.
She finds a brown dwarf star and notes down the position of the star.
She then switches to a visible light telescope, but is not able to see the star.

a) What does this tell the astronomer about the wavelength of light emitted by the star?

.... [1]

b) What conclusion can be drawn about the temperature of the star compared to the Sun?

.... [1]

c) Use the idea of black body radiation to explain your answers to **a)** and **b)**.

....

....

....

....

[3]

Total Marks .... / 8

An Introduction to Electricity

1. Circle the correct words to complete the sentences.

   Electric **current / charge** is the flow of electrical **charge / potential**.

   The **greater / smaller** the flow, the higher the **current / voltage**. [4]

2. The element in a set of hair straighteners has a 5A current running through it and a 230V potential difference across it.

   a) Write down the equation that links potential difference, current and resistance.

   Answer .... [1]

   b) Calculate the resistance of the element.

   Answer .... [2]

   c) Write down the equation that links charge, current and time.

   Answer .... [1]

   d) The straighteners are used for 2 minutes.

   Calculate the charge that flows in this time.

   Answer .... [2]

3. Circle the correct words to complete the sentences.

   Potential difference determines the amount of **energy / power** transferred by the charge as it passes through a component.

   The **greater / smaller** the potential difference, the higher the **current / voltage** that will flow. [3]

4. Write the name of the component that each circuit symbol represents.

   a) ⎯oo⎯ Answer .... [1]

   b) ⊣⊢⸱⸱⸱⊣⊢ Answer .... [1]

   c) ⎯▭⎯ Answer .... [1]

   d) Answer .... [1]

   Total Marks .... / 17

Circuits and Resistance

1 The circuit in **Figure 1** is used to measure the current and potential difference of various components.

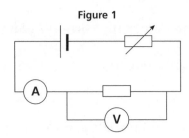

Figure 1

a) What is the purpose of the variable resistor?

....

.... **[2]**

b) Is the ammeter in **Figure 1** connected in series or parallel with the variable resistor?

Answer .... **[1]**

c) Is the voltmeter in **Figure 1** connected in series or parallel with the resistor?

Answer .... **[1]**

2 Draw **one** line from each component to the correct description.

Light dependent resistor (LDR)	Resistance decreases as temperature increases.
Thermistor	Resistance increases as the temperature increases.
Diode	Resistance decreases as light intensity increases.
Filament light	Has a very high resistance in one direction.

[3]

Total Marks .... / 7

Circuits and Power

1 This question is about a hairdryer that heats air and blows it out the front through a nozzle.

a) The hairdryer has an input power of 1600W.

  If a person takes two minutes to dry their hair, how much energy has been transferred?

  Answer _____ [3]

b) The hairdryer is 90% efficient. The remaining energy is output as sound.

  How much sound energy is produced in two minutes of use?

  Answer _____ [2]

2 **Figure 2** shows a series circuit with two resistors, **X** and **Y**.

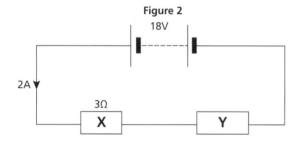

Figure 2

a) Calculate the potential difference across resistor **X**.

  Answer _____ [2]

b) Use your answer to part **a)** to work out the potential difference across component **Y**.

  Answer _____ [2]

c) Calculate the total resistance of the circuit.

  Answer _____ [2]

Total Marks _____ / 11

1 A battery is connected to an oscilloscope and the trace in **Figure 1** is produced.

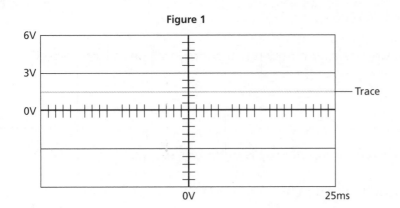

Figure 1

a) Use the trace to determine the potential difference of the battery and the type of current.

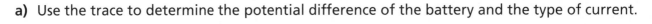

[2]

b) The battery is replaced by the mains supply and the trace in **Figure 2** is recorded by the oscilloscope.

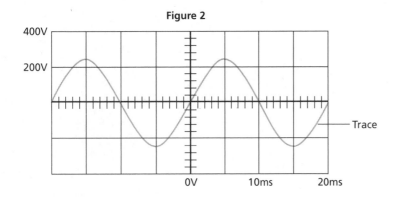

Figure 2

Use the trace to determine the potential difference and the type of current.

[2]

2 An appliance is switched off but the power cable is connected to the mains.

Explain how a live wire can still be dangerous.

[4]

Total Marks .... / 8

Electrical Energy in Devices

1 An electric blender transfers electrical energy into kinetic, heat and sound energy.

a) What is the useful energy output?   Answer .... [1]

b) What happens to the waste energy produced?

....

.... [2]

2 **Table 1** gives some information about an electric drill.

Table 1

Energy Input	
Useful Energy Output	
Wasted Energy	Heat and Sound
Power Rating	500W

a) Complete **Table 1** by adding the missing types of energy.   [2]

b) Which of the following statements about the energy from the drill is **incorrect**?
Tick **one** box.

It spreads out and becomes more difficult to use.   ☐

It disappears.   ☐

It makes the surroundings warmer.   ☐   [1]

c) How much energy does the drill use per second?   Answer .... [1]

3 The National Grid distributes electricity from power stations to consumers.
The voltage across the overhead cables of the National Grid is much higher than the output
voltage from the power station generators.

Explain how this achieved and why it is important.

....

....

....

....

.... [4]

Total Marks .... / 11

Static Electricity

1 Write down whether each of the statements about electric fields around objects is **true** or **false**.

a) The strength of the field depends on the size of the charge on the object.

Answer _____ [1]

b) The strength of the field depends on whether the charge is positive or negative.

Answer _____ [1]

c) The strength of the field depends on the material that the object is made from.

Answer _____ [1]

d) The strength of the field depends on the distance from the object.

Answer _____ [1]

2 a) A student rubs a Perspex rod with a woollen jumper and the Perspex rod becomes positively charged. Describe what has taken place to make this happen.

_____

_____

_____ [3]

b) **Figure 1** shows what happens when different charged rods interact.

Use the information in **Figure 1** to help you complete the sentence.

Figure 1

Perspex rod repels a perspex rod

Perspex rod attracts an ebonite rod

Like charges _____ and unlike charges _____ . [2]

3 Complete the sentences using words from the box.
Each word may be used once, twice or not at all.

charges	force	electric	magnetic	energy	charge	charged

When a _____ object is placed in an _____ field caused by another object,

it will experience a _____ .

The direction of the _____ depends on the _____ of the objects. [5]

Total Marks _____ / 14

Magnetism and Electromagnetism

1 The full name for the north pole of a magnet is the 'north-seeking pole'.

Explain what is meant by this.

....

.... [2]

2 The north pole of a permanent magnet is moved close to the north pole of another permanent magnet.

a) What would you expect to happen?

.... [1]

b) A piece of iron is moved close to the north pole of a permanent magnet.

What would you expect to happen?

.... [1]

3 **Figure 1** shows an electric bell.

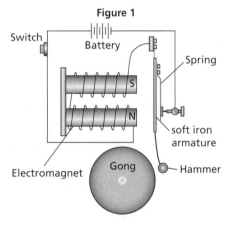

Describe what happens when the switch is pressed.

....

....

....

.... [5]

Total Marks .... / 9

The Motor Effect

1 **Figure 1** shows a loudspeaker.

Loudspeakers use the motor effect to convert electrical energy into sound energy.

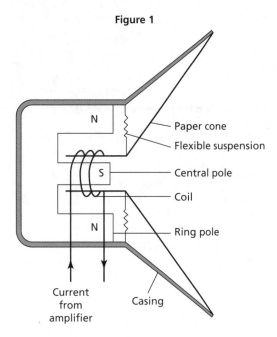

Figure 1

Explain how the input electrical signal is converted into sound.

....

....

....

....

....

....

....

.... [4]

2 A current carrying wire passes through a magnetic field at right-angles to the field and experiences a force.

The length of wire in the field is 5cm, the current is 2A and the magnetic field is 0.3mT.

a) Calculate the force on the wire.
Use the correct equation from the Physics Equation Sheet on page 215.

Answer .... [2]

b) The magnets are rearranged so that the current flowing in the wire is parallel to the field lines.

How will this affect the force on the wire?

.... [1]

Total Marks .... / 7

Induced Potential and Transformers

1 There are two main types of generators: one produces alternating current and the other produces direct current.

a) Which type of generator produces direct current?   Answer _____ [1]

b) A generator that produces a direct current can be used to power the lights on a bicycle. The rotation of the tyre is used to rotate the coil in the generator.
Using this device as the **only** method of powering a bicycle light is not considered safe.

Suggest why.

_____

_____ [2]

2 **Figure 1** shows how the output current from an alternator varies with time and the relative position of the magnet within the coil.

Figure 1

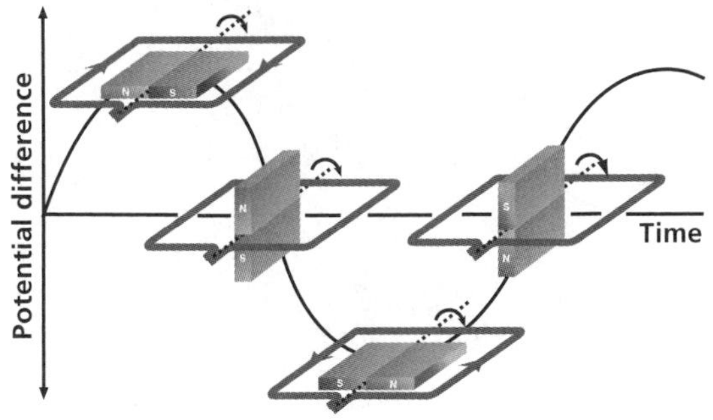

a) In terms of amplitude and frequency, what effect would doubling the speed of rotation have on the graph in **Figure 1**?

_____

_____

_____

_____ [3]

b) In terms of amplitude and frequency, what effect would doubling the strength of the magnet have on the graph in **Figure 1**?

_____

_____ [2]

c) The induced current creates its own magnetic field.

How will this magnetic field affect the force needed to rotate the magnet?

_____

_____ [2]

Total Marks _____ / 10

Particle Model of Matter

1 Heating a substance can cause it to change state from a solid to a liquid or from a liquid to a gas.

a) What is meant by 'specific latent heat of fusion'?

....

.... **[2]**

b) While a kettle boils, 0.012kg of water changes to steam.

Calculate the amount of energy required for this change.
Use the correct equation from the Physics Equation Sheet on page 215.
Specific latent heat of vaporisation of water = 2.3×10^6 J/kg

Answer .... **[2]**

2 The graph in **Figure 1** shows how temperature varies with time as a substance cools
The graph is **not** drawn to scale.

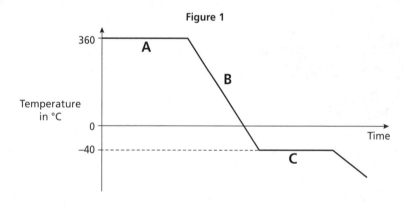

Figure 1

a) Explain what is happening to the substance in section **A** of the graph.

....

.... **[2]**

b) Explain what is happening to the substance in section **B** of the graph.

....

....

.... **[2]**

Total Marks .... / 8

Atoms and Isotopes

1 Atoms contain three types of particle.

a) Complete the table to show the relative charges of the subatomic particles.

Particle	Relative Charge
Electron	
Neutron	
Proton	

[3]

b) A neutral atom has no overall charge.

Explain why in terms of its particles.

....

....

.... [2]

c) Complete the sentences below.

An atom that loses or gains an electron becomes an ....

If it loses an electron, it has an overall .... charge. [2]

2 In the early part of the 20th century, Rutherford and Marsden investigated the paths taken by positively charged alpha particles into and out of a very thin piece of gold foil. **Figure 1** shows the paths of three alpha particles.

Explain the different paths, **A**, **B** and **C**, of the alpha particles.

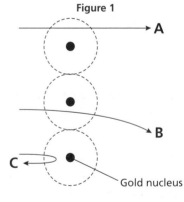

Figure 1

Gold nucleus

....

....

....

.... [3]

Total Marks .... / 10

Nuclear Radiation

1 Here is some information about potassium.

> **Potassium is a metallic element in Group 1 of the Periodic Table.**
> **It has an atomic number of 19.**
> **Its most common isotope is potassium-39, $^{39}_{19}\text{K}$.**
> **Another isotope, potassium-40, $^{40}_{19}\text{K}$, is a radioactive isotope.**

a) What is meant by 'radioactive isotope'?

.... [1]

b) During radioactive decay, atoms of potassium-40 change into atoms of a calcium-40.
Calcium-40 has an atomic number of 20 and a mass number of 40.

What type of radioactive decay has taken place?

Answer .... [1]

c) Potassium-39 does not undergo radioactive decay.

What does this tell us about potassium-39?

.... [1]

d) Sodium-24 is another radioactive isotope.
It decays by gamma emission.

Give the name of the element formed when this decay takes place.

Answer .... [1]

2 Give the unit that is used to measure the activity of a radioactive isotope.

Answer .... [1]

3 List the decay mechanisms, **alpha**, **beta** and **gamma**, in order of penetrating power.
Start with the most penetrative.

.... [1]

> **Total Marks** .... / 6

Using Radioactive Sources

1 Iodine is found naturally in the world and is essential to life.
It is used by the thyroid gland for the production of essential hormones.
Iodine-127 is not radioactive but Iodine-131 is.
Iodine-131 has as a half-life of 8 days.

a) During the Chernobyl nuclear disaster in 1986, an explosion caused a large quantity of the isotope iodine-131 to be released into the atmosphere.

Is iodine-131 from the disaster still a threat to us today?
Explain your answer.

....

....

.... [3]

b) Iodine-131 decays by beta emission and can be used for the treatment of thyroid cancer.

Explain why iodine-131 is suitable for this application.

....

....

....

.... [4]

c) A sample of iodine-131 has a count-rate of 256 counts per minute.

Work out the count-rate of the sample after 24 days.

Answer .... [2]

2 Radioactive isotopes can be used for medical tracers.

Explain how a medical tracer that is ingested in a drink can be used to look for blockages in the intestines.

....

....

.... [3]

Total Marks .... / 12

Fission and Fusion

1 The chart in **Figure 1** shows the sources of background radiation in Britain.

Figure 1

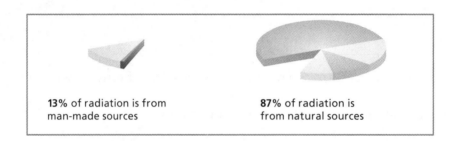

Radon gas
Released at surface of ground
from uranium in rocks and soil

Medical
Mainly
X-rays

From food

**Nuclear
industry**

Cosmic rays
From outer space and the Sun

Gamma rays
From rocks and soil
and building materials

13% of radiation is from
man-made sources

87% of radiation is
from natural sources

a) Give **two** sources of natural radioactivity from the chart.

.... [2]

b) Suggest how the chart might be used to reassure people that nuclear power is safe.

....

.... [1]

2 Complete the nuclear equations below by filling in the blanks.

a) Alpha decay: $^{219}Ra \rightarrow\ ^{\ \ \ }_{84}Po +\ ^{\ \ \ }_{\ \ \ }He$ [2]

b) Beta decay: $^{234}_{\ \ \ }Th \rightarrow\ ^{\ \ \ }_{91}Pa +\ ^{\ \ \ }_{\ \ \ }e$ [2]

Total Marks .... / 7

Stars and the Solar System

1 Stars are formed from massive clouds of dust and gases in space.

a) What force pulls the clouds of dust and gas together to form stars?

Answer _____ [1]

b) Once formed, a star can have a stable life for billions of years.

Describe the **two** main forces at work in the star during this period of stability.

_____

_____ [2]

c) What happens to the star once this stable period is over?

_____

_____ [2]

d) Suggest what might happen to a planet close to the star at this time?

_____ [1]

e) What happens in the final stages of the lifecycle of the largest stars?

_____ [1]

2 At the very high temperatures in the Sun, hydrogen is converted into helium.
It takes four hydrogen nuclei to produce one helium nucleus.
Table 1 shows the relative masses of hydrogen and helium nuclei.

Table 1

Nucleus	Relative Mass
Hydrogen	1.007825
Helium	4.0037

a) Use the data in **Table 1** to calculate by how much the mass of the Sun is reduced every time hydrogen is converted to helium.

Answer _____ [2]

b) Explain how the Sun is able to radiate huge amounts of energy for billions of years but will eventually become a red giant.

_____

_____ [2]

Total Marks _____ / 11

Orbital Motion and Red-Shift

1 A man-made satellite orbits the Earth.
The satellite experiences a resultant force directed towards the centre of the orbit.

a) What provides the force on the satellite?   Answer .... [1]

b) Why does this force not cause the satellite to change speed?

....

.... [2]

c) What effect, if any, would changing the speed of a satellite in a stable orbit have?

.... [1]

d) In which direction is the instantaneous velocity of a satellite in orbit?

                                                           Answer .... [1]

2 Red-shift is one of the pieces of evidence that led scientists to propose the 'Big Bang' theory.

a) Describe the Big Bang theory.

....

.... [2]

b) What is meant by red-shift?

....

.... [2]

c) Explain how red-shift provides evidence for the Big Bang theory.

....

....

....

.... [2]

Total Marks .... / 11

Collins

GCSE
PHYSICS
Paper 1 Higher Tier

H

Materials

Time allowed: 1 hour 45 minutes

For this paper you must have:

- a ruler
- a calculator
- the Physics Equation Sheet (page 215).

Instructions

- Answer **all** questions in the spaces provided.
- Do all rough work on the page. Cross through any work you do not want to be marked.

Information

- There are **100** marks available on this paper.
- The marks for each question are shown in brackets [ ].
- You are expected to use a calculator where appropriate.
- You are reminded of the need for good English and clear presentation in your answers.

Advice

- In all calculations, show clearly how you work out your answer.

01 **Figure 1** shows what happens to each 100 joules of energy from coal that is burned in a power station.

Figure 1

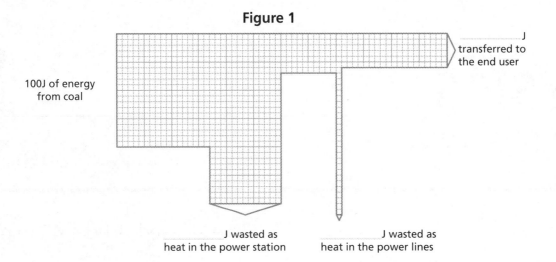

100J of energy from coal

J transferred to the end user

_____ J wasted as heat in the power station

_____ J wasted as heat in the power lines

01.1 Add the missing figures to the diagram. **[3 marks]**

01.2 For the same cost, the electricity company could install new power lines that only waste half as much energy as the old ones OR use a quarter of the heat wasted at the power station to heat schools in a nearby town.

Which of these two things do you think they should do?
Give a reason for your answer.

_____

_____

_____

_____

[4 marks]

01.3 Calculate the efficiency of the coal powered station in **Figure 1**.

Efficiency = _____ %   **[1 mark]**

02   A gas burner is used to heat some water in a pan.

By the time the water starts to boil:

- 60% of the energy released has been transferred to the water
- 20% of the energy released has been transferred to the surrounding air
- 13% of the energy released has been transferred to the pan
- 7% of the energy released has been transferred to the gas burner itself.

02.1   Some of the energy released by the burning gas is wasted.

What happens to this wasted energy?

_____

_____   [2 marks]

02.2   What percentage of the energy from the gas is wasted?

Percentage = _____ %   [1 mark]

02.3   How efficient is the gas burner at heating water?

Efficiency = _____ %   [1 mark]

Turn over for the next question

03 A book weighs 6 newtons.

A librarian picks up the book from the ground and puts it on a shelf that is 2 metres high.

03.1 Calculate the work done on the book.

Work done = _____ J   [2 marks]

03.2 The next person to take the book from the shelf accidentally drops it.
The book falls 2 m to the ground.

Calculate how much gravitational energy it loses as it falls.
The gravitational field strength is = 10 N/kg.

Answer = _____ J   [2 marks]

03.3 All of the book's gravitational energy is converted to kinetic energy when it falls.

Calculate the velocity with which the book hits the floor.

Velocity = _____ m/s   [3 marks]

04 Three different types of polystyrene were suggested as thermal insulators. Their densities are shown in **Table 1**.

Table 1

Type of Polystyrene	Density (kg/m³)
A	12
B	18
C	24

04.1 Predict which material will act as the best thermal insulator. Explain your prediction.

....

....

....

....

[2 marks]

04.2 Describe an experiment that could be used to test your prediction, including the variables which need to be controlled.

....

....

....

....

....

....

....

....

....

[6 marks]

04.3 Give one source of inaccuracy in this investigation and a method of improving it.

....

....

....

[2 marks]

Question 4 continues on the next page

04.4 Give one hazard of this investigation and a suitable control method.

[2 marks]

05 A computer is set up to produce a graph of the current through an electric lamp for the first few milliseconds after it is switched on.

The graph is shown in **Figure 2**.

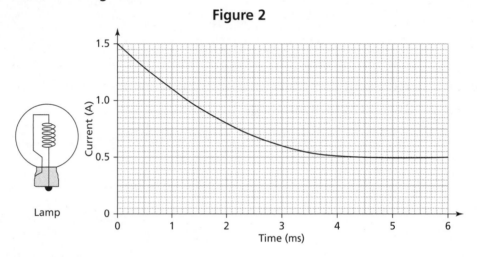

Figure 2

05.1 Use the graph to describe how the current through the lamp changes when it is switched on.

[3 marks]

05.2 There is a constant 12 V potential difference across the lamp.

What conclusion can be drawn about how the resistance of the lamp changes in the first few milliseconds after it is switched on?

You should state the resistance at different points to support your conclusion.

[3 marks]

06 A student carries out an experiment to investigate the current through component **X**.

A circuit is set up as shown in **Figure 3**.

The current is measured when different voltages are applied across component **X**.

Figure 3

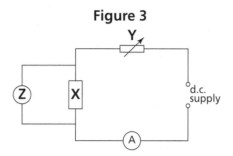

06.1 Name the components labelled **Y** and **Z** in the circuit.

Y = ...

Z = ... [2 marks]

06.2 What is the role of component **Y** in the circuit?

.... [1 mark]

Table 2 shows the measurements obtained in this experiment.

Table 2

Voltage (V)	−0.6	−0.4	−0.2	0	0.2	0.4	0.6	0.8
Current (mA)	0	0	0	0	0	50	100	150

06.3 Name the independent variable in this experiment.

Independent variable = ... [1 mark]

06.4 Name the dependent variable in this experiment.

Dependent variable = ... [1 mark]

Question 6 continues on the next page

06.5   Plot a graph on the axes in **Figure 4** using the data from **Table 1**.

Figure 4

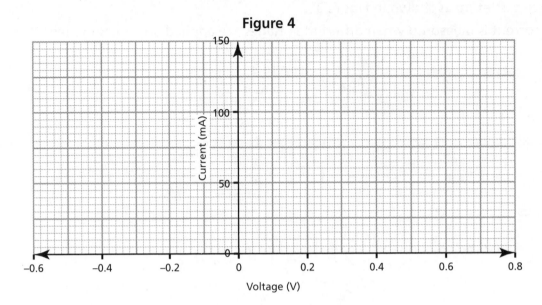

Current (mA)

Voltage (V)

[3 marks]

06.6   The student looks at their measurements and decides that there are no anomalous results.

Are they correct?
You must explain your answer.

_____

_____   [1 mark]

06.7   Use the shape of the graph to name component **X**.

Component **X** = _____   [1 mark]

07 A student carries out an experiment to investigate static charge.
Here is the method that they use:

1. Take a polythene rod, hold it at its centre and rub both ends with a cloth.
2. Suspend the rod, without touching the ends.
3. Take a Perspex rod and rub it with another cloth.
4. Without touching the ends of the Perspex rod, bring each end of the Perspex rod close to each end of the polythene rod.
5. Make notes on what is observed.

Figure 5 shows how the apparatus is set up.

Figure 5

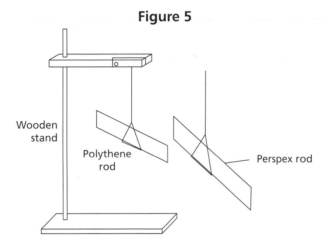

07.1 When the Perspex rod is brought close to the polythene rod they attract each other.

Explain why they attract each other.

....

.... **[2 marks]**

07.2 Describe what will happen when the Perspex rod is reversed and the opposite end is brought close to the polythene rod.

.... **[1 mark]**

Question 7 continues on the next page

The experiment is repeated with two polythene rods.

07.3 Describe what will happen when the end of one rod is brought near to the end of the other.

.... **[1 mark]**

07.4 Explain your answer to **07.3**.

....

.... **[2 marks]**

07.5 Explain, in terms of electron movement, what happens as the rods are rubbed with the cloths.

....

....

....

.... **[3 marks]**

08   An adaptor can be used to connect up to four appliances in parallel to one 230 V mains socket.

Table 3 gives a list of appliances and the current that each one draws from a mains socket.

Table 3

Appliance	Current
Computer	1 A
Hairdryer	4 A
Heater	8 A
Iron	6 A
Television	2 A

08.1   What current will flow to the adaptor when the television, computer and hairdryer are plugged into the adaptor?

Current = _____ A   [1 mark]

08.2   Calculate the electrical power used when the television, computer and hairdryer are plugged into the adaptor.

Power = _____ W   [2 marks]

Question 8 continues on the next page

08.3   Explain why it could be dangerous to plug all the devices into the adaptor
at the same time.

_____

_____

_____

_____   **[2 marks]**

08.4   Use your answers to **08.1** and **08.2** to calculate the combined resistance when
the television, computer and hairdryer are all plugged in and operating.

Resistance = _____ Ω   **[2 marks]**

08.5   The television, computer and hairdryer are left running for 5 minutes.

Calculate the charge that flows through the adaptor in this time.

Charge = _____ C   **[2 marks]**

09 There are many isotopes of the element strontium (Sr).

09.1 What do the nuclei of different strontium isotopes have in common?

[1 mark]

09.2 The isotope strontium-90 is produced inside nuclear power stations from the fission of uranium-235.

What happens during the process of nuclear fission?

[2 marks]

09.3 When the nucleus of a strontium-90 atom decays, it emits radiation and changes into a nucleus of yttrium-90.

$$^{90}_{38}\text{Sr} \rightarrow ^{90}_{39}\text{Y} + \text{Radiation}$$

What type of decay is this?

Answer _____ [1 mark]

09.4 Give a reason for your answer to **09.3**.

[1 mark]

Question 9 continues on the next page

Strontium-90 has a half-life of 30 years.

09.5 What is meant by the term 'half-life'?

.... **[1 mark]**

09.6 After formation in the nuclear reactor, strontium-90 is stored as radioactive waste.

For how many years does strontium-90 have to be stored before its radioactivity has fallen to $\frac{1}{8}$ of its original level?

Answer .... **[2 marks]**

09.7 In a fission reaction strontium-89 is also produced.
Strontium-89 has a half-life of 50 days.

Explain why this makes it both a more and a less hazardous waste product.

....

....

....

.... **[2 marks]**

10   A car that is moving has kinetic energy. The faster a car goes, the more kinetic energy it has.

   10.1   The kinetic energy of a car was 472 500 J when travelling at 30 m/s.

         Calculate the total mass of the car.
         Give the unit.

                                                      Mass = ...   [4 marks]

There is a government road safety campaign to reduce the speed at which people drive in
residential areas. It uses the slogan 'Kill your speed, not a child'.
The scientific reason for this is that kinetic energy is transferred from the vehicle to the person
it knocks down.

   10.2   A bus and car are travelling at the same speed.
         The bus is likely to cause more harm to a person who is knocked down than the van would.

         Explain why.

         ....

         ....

         ....   [2 marks]

   10.3   A car and its passengers have a total mass of 1200 kg.
         The car is travelling at 8 m/s.

         Calculate the increase in kinetic energy when the car increases its speed to 14 m/s.

                                            Increase .... J   [3 marks]

   10.4   Explain why the increase in kinetic energy is much greater than the increase in speed.

         ....

         ....   [1 mark]

                        Turn over for the next question

11 In order to jump over the bar, a high jumper must raise his mass above the ground by 1.25 m. The high jumper has a mass of 65 kg.

The gravitational field strength is 10 N/kg.

11.1 The high jumper just clears the bar.

Calculate the gain in his gravitational potential energy.

Gain = _____ J   **[2 marks]**

11.2 Calculate the minimum vertical speed the high jumper must reach in order to jump over the bar.

Use your answer to **11.1** and the formula for kinetic energy.

Minimum vertical speed = _____ m/s   **[3 marks]**

12 The circuit diagram in **Figure 6** shows a circuit used to supply electricity for car headlights.

Figure 6

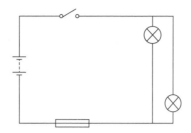

The current through the filament of one car headlight is 2 A.
The potential difference supplied by the battery is 12 V.

12.1 What is the potential difference across each headlight?

Potential difference = ... V **[1 mark]**

12.2 Work out the total current through the battery.

Current = ... A **[1 mark]**

12.3 Which of the following ratings should be used for the fuse in the circuit?

Tick **one** box.

3 A   ☐

5 A   ☐

10 A   ☐

13 A   ☐   **[1 mark]**

12.4 Calculate the resistance of each headlight filament when in use.

Resistance = ... Ω **[2 marks]**

Question 12 continues on the next page

12.5 How does the total resistance of the circuit compare to the resistance of each individual bulb?

_____ [1 mark]

12.6 Calculate the power supplied to each of the two headlights of the car.

Power = _____ W [2 marks]

12.7 The fully charged car battery can deliver 96 kJ of energy at 12 V.

How long can the battery keep both the headlights fully on?

Length of time = _____ s [2 marks]

END OF QUESTIONS

Collins

GCSE
PHYSICS
Paper 2 Higher Tier

H

Materials

Time allowed: 1 hour 45 minutes

For this paper you must have:

- a ruler
- a calculator
- a protractor
- the Physics Equation Sheet (page 215).

Instructions

- Answer **all** questions in the spaces provided.
- Do all rough work on the page. Cross through any work you do not want to be marked.

Information

- There are **100** marks available on this paper.
- The marks for each question are shown in brackets [ ].
- You are expected to use a calculator where appropriate.
- You are reminded of the need for good English and clear presentation in your answers.

Advice

- In all calculations, show clearly how you work out your answer.

Practice Exam Paper 2

01 A student used a lever system to investigate how the force of attraction between a coil and an iron rocker varied with the current in the coil.
She supported a coil vertically and connected it to an electrical circuit as shown in **Figure 1**.
The weight of the iron rocker is negligible.

Figure 1

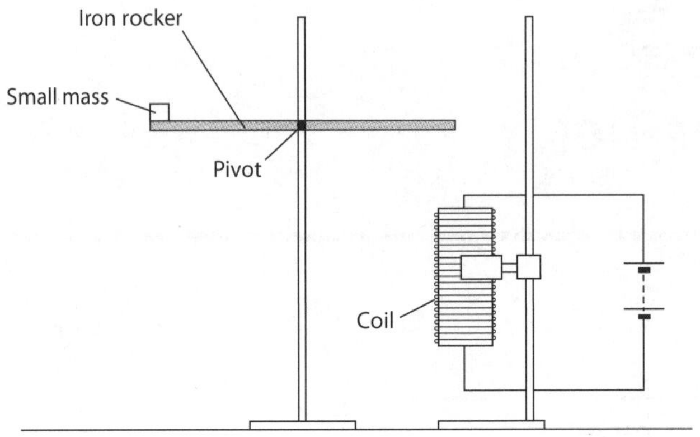

01.1 Why is it important that the rocker in this experiment is made of iron?

[1 mark]

01.2 The student put a small mass on the end of the rocker and adjusted the current in the coil until the rocker balanced.

To keep the rocker balanced, how will the current through the coil need to change as the size of the mass is increased?

[1 mark]

01.3 Explain your answer to **01.2**.

[2 marks]

01.4 A second student set up the same experiment and put an iron core inside the coil.

How will this affect the size of the mass that can be balanced?
You must explain your answer.

_____

_____

_____ **[2 marks]**

01.5 The student decides to use the original electromagnet and lever system as a force multiplier.
By adjusting the position of the iron bar and the pivot it can be used to lift a larger mass, as shown in **Figure 2**.

Figure 2

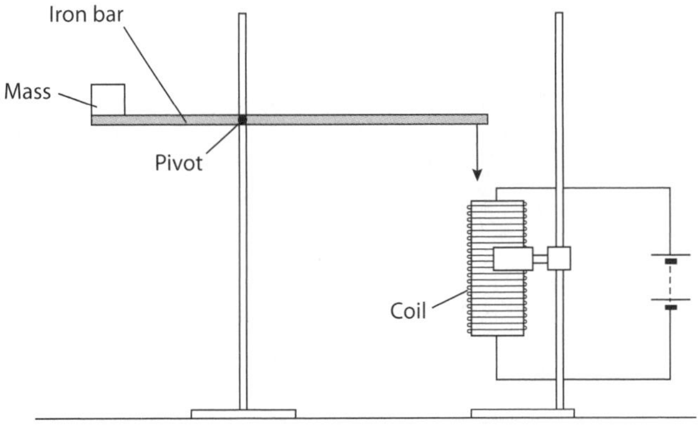

Which of these actions would allow the magnet to lift a heavier mass?

Tick **two** boxes.

Moving the mass closer to the pivot ☐

Moving the magnet closer to the pivot ☐

Reversing the direction of the magnet ☐

Increasing the number of batteries ☐ **[2 marks]**

Question 1 continues on the next page

01.6 The mass in **Figure 2** has a weight of 50 N.
It is positioned 5 cm from the pivot.

Calculate the turning moment created by the weight.

Turning moment = ... **Nm** [2 marks]

01.7 The electromagnet applies a force 20 cm from the pivot.

Use your answer to **01.6** to work out the force needed to exactly balance the weight.

Force = ... **N** [2 marks]

02 A group of students investigate circular motion.
They swing a bung attached to a string around in circle.
The string is attached to a force meter, which measures
the centripetal force.

Figure 3

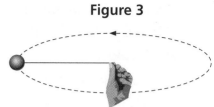

The students record how the reading on the force meter
changes to determine how the force affects the speed of the orbit.

02.1 In which direction does the centripetal force act on the rubber bung?

.... **[1 mark]**

02.2 In this experiment, what provides the centripetal force?

.... **[1 mark]**

02.3 One student swung the rubber bung around in a circle at constant speed.
A second student timed how long it took the rubber bung to complete 10 rotations.

Give **two** variables that are important to control in this experiment.

....

.... **[2 marks]**

02.4 The Moon orbits the Earth in a circular path.

direction	resistance	speed	velocity

Use words from the box to complete the sentences.
You may use each word once, more than once or not at all.

The Moon's .... is constant but its .... changes.

This is because its .... changes. **[1 mark]**

02.5 What force provides the centripetal force needed to keep the Moon in its orbit
around the Earth?

.... **[1 mark]**

Turn over for the next question

03 **Figure 4** shows a transformer.

There is a 50 Hz (a.c.) supply connected to 10 turns of insulated wire wrapped around one side of the iron core.

A voltmeter is connected to 5 turns wrapped around the other side of the iron core.

Figure 4

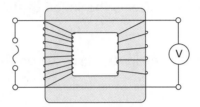

Table 1 shows values for the potential difference (p.d.) of the supply and the voltmeter reading.

Table 1

Potential Difference of the Supply (V)	Voltmeter Reading (V)
6.4	3.2
3.2	....
....	6.4

03.1 Complete **Table 1**. **[2 marks]**

03.2 Explain in terms of magnetic fields how a transformer works.

....

....

....

....

....

....

[4 marks]

04 Some students fill an empty plastic bottle with water.
The weight of the water in the bottle is 20 N.
The cross-sectional area of the bottom of the bottle is 0.006 m².

04.1 Calculate the pressure of the water on the bottom of the bottle.
Give your answer to 2 significant figures.

Pressure = ... N/m²   **[3 marks]**

The students made three holes in the bottle along a vertical line: hole A is at the bottom,
B is in the middle, and C is near the top.

04.2 From which hole will the water come out at the slowest speed?

Answer = ... **[1 mark]**

04.3 Explain your answer to **4.2**.

....

.... **[2 marks]**

05 **Figure 5** shows an electromagnetic switch used in a starter motor circuit for a car.

Figure 5

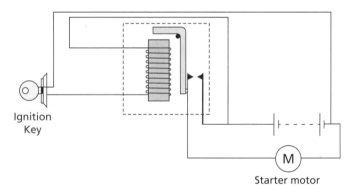

Explain how turning the ignition key makes a current flow in the starter motor.

....

....

....

.... **[3 marks]**

Turn over for the next question

06   All stars go through a lifecycle.
About 90% of all stars, including the Sun, are currently in the main sequence period of their lifecycle.

06.1   Explain briefly how stars like the Sun are thought to have been formed.

....

.... **[2 marks]**

06.2   Explain why stars are stable during the main sequence period of their lifecycle.

....

.... **[1 mark]**

06.3   The length of time that a star remains in the main sequence of its lifecycle depends on it size and temperature.
Hotter larger stars do not stay in the main sequence for as long as cooler smaller stars.

What does this suggest about the link between the rate of fusion and the size of the star?

....

.... **[2 marks]**

06.4   In 1929, the astronomer Edwin Hubble observed that the light from galaxies moving away from the Earth showed a red-shift.
Red-shift provides evidence for the theory that the universe began from one very small initial point.

What name is given to the theory that the universe began in this way?

Answer = ... **[1 mark]**

06.5   Although the early universe contained only hydrogen, it now contains many different elements.

Describe how the different elements were formed.

....

....

....

.... **[2 marks]**

07   **Figure 6** shows a coil and a magnet.
      An ammeter is connected to the coil.

      The ammeter has a centre zero scale so that values
      of current going in either direction through the
      coil can be measured.

Figure 6

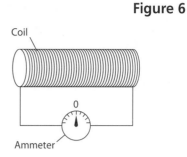
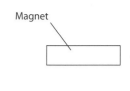

07.1   When the magnet moves into the coil, the needle on the ammeter moves.

       Explain why this happens.

       ....

       ....

       ....

       ....
                                                                                                    [4 marks]

       Table 2 lists other possible ways of moving the magnet in relation to the coil.

                                       Table 2

Movement of the Magnet	What Happens to the Ammeter Reading?
Hold the magnet stationary within the coil.	
Move the magnet quickly towards the coil.	
Reverse the magnet and move it slowly towards the coil.	

07.2   Complete **Table 2** by writing down the effect of each action on the
       ammeter reading.

                                                                                                    [3 marks]

07.3   The current induced in the solenoid creates a magnetic field.
       The north pole of the magnet is moved towards the solenoid.

       What pole will be induced on the end of the solenoid closest to the magnet?

                                       Answer = ...   **[1 mark]**

                   Turn over for the next question

08   A mobile phone uses a transformer to recharge the battery.
The transformer is connected to a 230 V mains supply and has a 4.6 V output.

08.1   Explain how you know that this is a step-down transformer?

....

.... **[1 mark]**

08.2   Describe the construction of a step-down transformer.
You may add to **Figure 7** to help with your answer.

Figure 7

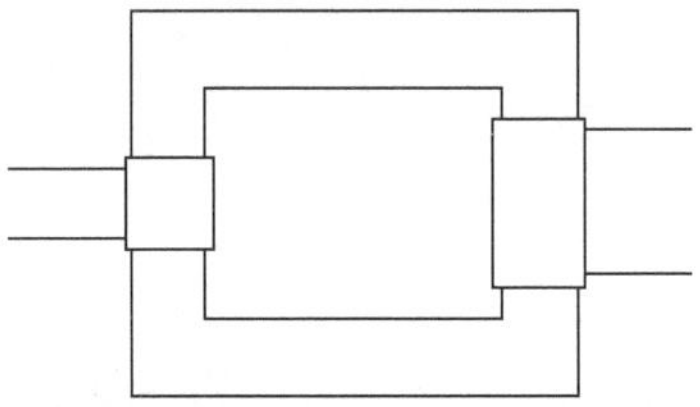

....

....

....

.... **[4 marks]**

08.3   The transformer has 2000 turns on the primary coil.

Calculate the number of turns on the secondary coil.
Use the correct equation from the Physics Equation Sheet on page 215.

Answer = ... **[2 marks]**

09 Human ears can detect a range of sound frequencies.

09.1 Complete the sentence.

The range of human hearing is from about _____ Hz to _____ Hz.   **[2 marks]**

09.2 What is ultrasound?

_____   **[1 mark]**

09.3 The speed of an ultrasound wave in soft tissue in the human body is 1500 m/s. The frequency of the wave is 2.0×10^6 Hz.

Calculate the wavelength of the ultrasound wave.

Wavelength = _____ m   **[2 marks]**

09.4 Describe how ultrasound can be used to find out if a patient is suffering from kidney stones (a build-up of hard mineral deposits in the kidneys).

_____

_____

_____

_____   **[2 marks]**

Turn over for the next question

10 **Figure 8** shows how a convex lens forms an image of an object.
    It is **not** drawn to scale.

Figure 8

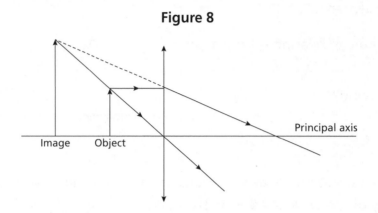

10.1 Which of these words can be used to describe the image?
      Draw a ring around **two** words.

    diminished   **inverted**   **magnified**   **real**   **upright**   **[2 marks]**

10.2 The object is 2 cm tall and the image is 6 cm tall.

Work out the magnification of the lens.
Use the correct equation from the Physics Equation Sheet on page 215.

Magnification = _____ **[1 mark]**

11 A ripple tank is a piece of lab equipment that can be used to investigate the properties of water waves.

11.1 Distinguish between the amplitude and the wavelength of a wave.

....

....

....

.... **[2 marks]**

11.2 Explain how frequency, wavelength and speed of a wave can be measured using a ripple tank. Your answer should consider any cause of inaccuracy in the data.

....

....

....

....

....

....

....

....

....

.... **[6 marks]**

Question 11 continues on the next page

11.3   Explain the importance of controlling the depth of the water in the ripple tank.

[2 marks]

11.4   Describe a hazard in this investigation and how you would control it.

[2 marks]

12   **Figure 9** shows a lens used as a magnifying glass.

The position of the eye is shown.

The arrow shows the size and position of an object at point **O**.

Use a ruler to accurately construct the position of the image on **Figure 9**.

You should show how you construct your ray diagram and how light appears to come from this image to enter the eye.

Figure 9

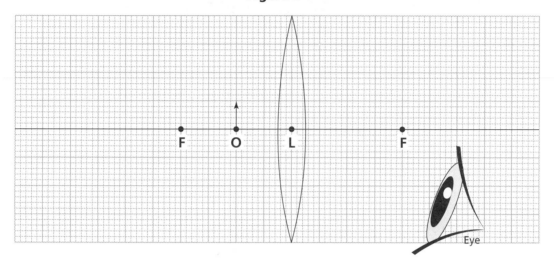

[4 marks]

Turn over for the next question

13 Radio waves, ultraviolet waves, visible light and X-rays are all types of electromagnetic radiation.

13.1 Choose wavelengths from the list below to complete **Table 3**.

3×10^{-8} m 1×10^{-11} m 5×10^{-7} m   1500 m

Table 3

Type of Radiation	Wavelength (m)
Radio waves	
Ultraviolet waves	
Visible light	
X-rays	

[3 marks]

13.2 Radio waves can be used to control remote control cars.

Calculate the frequency of radio waves of wavelength 300 m.
(The velocity of electromagnetic waves is 3×10^8 m/s.)

Frequency = _____ **Hz** [3 marks]

The graph in **Figure 10** shows the speed of a remote-controlled vehicle during a race.

Figure 10

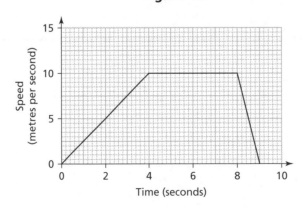

13.3 Calculate the acceleration during the first four seconds.

Acceleration = ... m/s²   **[3 marks]**

13.4 What is the maximum speed reached by the vehicle?

Maximum speed = ... m/s   **[1 mark]**

13.5 How far does the vehicle travel between 4 and 8 seconds?

Distance = ... m   **[2 marks]**

13.6 At the finish line, a thick wall of rubber foam slows the vehicle down at a rate of 25 m/s². The vehicle has a mass of 1.5 kg.

Calculate the average force of the rubber foam on the car.

Average force = ... N   **[2 marks]**

Turn over for the next question

14   When a gun is fired, a very large force acts on the bullet for a very short time.
     An average force of 4000 newtons acts for 0.004 seconds on a bullet of mass 50 g.

   14.1   Calculate the momentum gained by the bullet.
          Use the correct equation from the Physics Equation Sheet on page 215.

                                        Momentum = ... kg m/s   [2 marks]

   14.2   Calculate the speed of the bullet.

                                        Speed = ... m/s   [2 marks]

                              END OF QUESTIONS

Physics Equation Sheet

1	pressure due to a column of liquid = height of column × density of liquid × gravitational field strength (g)	$p = h\rho g$
2	(final velocity)^2 − (initial velocity)^2 = 2 × acceleration × distance	$v^2 - u^2 = 2as$
3	force = $\dfrac{\text{change in momentum}}{\text{time taken}}$	$F = \dfrac{m\Delta v}{\Delta t}$
4	elastic potential energy = 0.5 × spring constant × (extension)^2	$E_e = \frac{1}{2}ke^2$
5	change in thermal energy = mass × specific heat capacity × temperature change	$\Delta E = mc\Delta\theta$
6	period = $\dfrac{1}{\text{frequency}}$	
7	magnification = $\dfrac{\text{image height}}{\text{object height}}$	
8	force on a conductor (at right-angles to a magnetic field) carrying a current = magnetic flux density × current × length	$F = BIl$
9	thermal energy for a change of state = mass × specific latent heat	$E = mL$
10	$\dfrac{\text{potential difference across primary coil}}{\text{potential difference across secondary coil}} = \dfrac{\text{number of turns in primary coil}}{\text{number of turns in secondary coil}}$	$\dfrac{V_p}{V_s} = \dfrac{n_p}{n_s}$
11	potential difference across primary coil × current in primary coil = potential difference across secondary coil × current in secondary coil	$V_s I_s = V_p I_p$
12	For gases: pressure × volume = constant	$pV = constant$

Answers

Topic-Based Questions

Page 148 Forces – An Introduction

1. a) Weight / gravity [1]; non-contact [1]; air resistance (or drag) [1]; contact [1]
    b) weight = mass × gravitational field strength / $W = mg$ [1]; weight = 120 000 × 10 = 1 200 000N [1]
    c) A correctly drawn horizontal arrow [1]; a vertical arrow that is half the length of the horizontal arrow [1]; and a diagonal arrow showing the total resultant force [1]

Resultant Force

2. a) Friction [1]
    b) Friction is a contact force [1]; lifting the glider off the track means there is no contact, so no friction [1]

Page 149 Forces in Action

1. a) Remove the weights [1]; if it returns to its original shape it was behaving elastically [1]
    b) i) 8.0 (cm) [1]
        ii) The extension appears to be linear [1]; and is increasing by 4cm each time [1]
2. a) 200 × 1.6 [1]; = 320N [1]
    b) Tree moment = man moment [1]; 320 = 0.2 × F [1]; $F = \frac{320}{0.2}$ = 1600N [1]

Page 150 Pressure and Pressure Differences

1. a) pressure = $\frac{\text{force normal to surface}}{\text{area of that surface}}$
        / $p = \frac{F}{A}$ [1]
    b) The man presses down on a small area [1]; this increases the pressure of the liquid [1]; at the car the pressure acts on a bigger area and $F = pA$ so the force is increased [1]
2. a) The balloon is less dense than the surrounding air [1]
    b) At a higher altitude the air is less dense [1]; this means the force of upthrust on the balloon falls as it rises [1]; eventually upthrust and weight are balanced, so there is no resultant vertical force on the balloon [1]

Page 151 Forces and Motion

1. a) Zero [1]

> Displacement is the distance from the start point. The person returns home – it is a circular journey – so the total displacement at the end of their journey is zero.

b) B [1]
c) Stationary [1]
d) A is travelling away from home [1]; D is travelling in the opposite direction (back towards home) [1]; D is travelling slightly faster than A [1]

2. a) speed = $\frac{\text{distance travelled}}{\text{time}}$ /
        $v = \frac{s}{t}$ [1]; = $\frac{180}{6}$ = 30m/s [1]
    b) 50 – 40 = 10m/s [1]
    c) 50 + 40 = 90m/s [1]

> The velocity changes from 50m/s to the east to 40m/s to the west.

Page 152 Forces and Acceleration

1. a) The mass of the system [1]
    b) The force accelerating the trolley (provided by the hanging masses) [1]
    c) The acceleration of the trolley [1]

> The independent variable is the one deliberately changed and the dependent variable is the one being measured.

2. a) acceleration = $\frac{\text{change in velocity}}{\text{time taken}}$ /
        $a = \frac{\Delta v}{t}$ [1]; acceleration = $\frac{16 - 4}{8}$ = 1.5 [1]; m/s² [1]
    b) resultant force = mass × acceleration / $F = ma$ [1]; force = 68 × 1.5 = 102N [1]
3. a) acceleration = $\frac{33}{11}$ [1]; = 3m/s² [1]
    b) force = 950 × 3 [1]; = 2850N [1]

Page 153 Terminal Velocity and Momentum

1. They could change position e.g. dive head first [1]; to become more aerodynamic / reduce air resistance [1]
2. a) force = $\frac{\text{change in momentum}}{\text{time taken}}$ /
        $F = \frac{m\Delta v}{\Delta t}$, 6 700 000 = (5000 × v) – (5000 × 0) [1];
        $v = \frac{6\ 700\ 000}{5000}$ [1]; = 1340m/s [1]

> change in momentum = final momentum – initial momentum = $mv - mu$

b) mass burned = 5000 × 5 × 10 = 250 000kg [1]; momentum given to fuel = 250 000 × 1340 = 335 000 000kg m/s [1]; spacecraft gains 335 000 000kg m/s [1]

> Momentum is conserved, so the momentum gained by the rocket is equal to the momentum given to the fuel.

c) momentum = mass × velocity / $p = mv$, 335 000 000 = 3 000 000 × v [1]; $v = \frac{335\ 000\ 000}{3\ 000\ 000}$ = 111.67m/s [1]
d) As the rocket burns fuel, it becomes lighter [1]; resultant force = mass × acceleration / $F = ma$, so acceleration (a) increases [1]; at a higher altitude, there is less air resistance [1]; so the resultant force increases [1]

Page 154 Stopping and Braking

1. a) 1.4 seconds [1]
    b) 1.4 × 15 [1]; = 21m [1]
    c) 4 – 1.4 = 2.6 seconds [1]
    d) momentum = mass × velocity / $p = mv$ [1]; 1300 × 15 = 19 500kg m/s [1]
    e) force = rate of change of momentum [1]; force = $\frac{19\ 500}{2.6}$ [1]; = 7500N [1]
    f) A correctly drawn graph line that starts horizontally at 15m/s [1]; then starts sloping downwards between 0.2s and 0.8s [1]; and has the same gradient on the downslope as the original line [1]

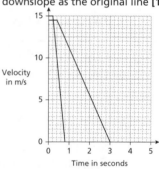

g) **Any two of:** The down slope would start at the same point [1]; but have a shallower gradient [1]; and take a longer total time to stop [1]

Page 155 Energy Stores and Transfers

1. a) 100 – 20 = 80 degree change [1]; energy = 2 × 4200 × 80 [1]; = 672 000J [1]
    b) 672 000 = 2.5 × 4200 × temp change [1]; temp change = $\frac{672\ 000}{(2.5 \times 4200)}$ = 64°C [1]; final temp = 20 + 64 = 84°C [1]
2. a) gravitational potential energy = mass × gravitational field strength × height / $E_p = mgh$ [1]; E_p = 0.1 × 10 × 0.05 [1]; = 0.05J [1]
    b) kinetic energy = 0.5 × mass × (speed)² / $E_k = \frac{1}{2}mv^2$ [1]; 0.05 = 0.5 × 0.1 × v^2 [1]; $v^2 = \frac{0.05}{(0.5 \times 0.1)}$ = 1, v = 1m/s [1]

Page 156 Energy Transfers and Resources

1. light [1]; electrical / thermal [1]; heat [1]
2. a) Start temperature of water [1]; thickness of fleece [1]
   b) Fleece M [1]; because it cools the slowest, so insulates the best [1]

Page 157 Waves and Wave Properties

1. a) Half a wave per second [1] (Accept: 1 wave every 2 seconds)
   b) wave speed = frequency × wavelength / $v = f\lambda$ [1]
   c) speed = $\dfrac{\text{distance}}{\text{time}}$ / $v = \dfrac{s}{t}$, $v = \dfrac{50}{10}$ [1]; = 5m/s [1]
   d) $v = f\lambda$, $\lambda = \dfrac{5}{0.5}$ [1]; = 10m [1]
   e) They will go slower [1]
2. In longitudinal waves, the particles oscillate [1]; parallel to the direction of energy transfer / wave motion [1]; in transverse waves, the oscillation is at right-angles to the direction of energy transfer / wave motion [1]

Page 158 Reflection, Refraction and Sound

1. Can be shown on the diagram to help explain but must include 'refraction' in the answer for full marks, e.g. Light rays from the pin [1]; are refracted when they leave the water [1]; away from the normal and into the eye [1]

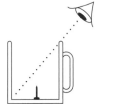

2. a) Two lines with arrows drawn to show reflection[1]; with the angle of incidence equal to the angle of reflection [1]

Sheet of metal

Loudspeaker

Sound meter

   b) Will make it quieter [1]; because the carpet absorbs the sound energy [1]

Page 159 Waves for Detection and Exploration

1. distance travelled = speed × time / $s = vt$ [1]; $s = 1600 \times 0.8 = 1280$m [1]; $\dfrac{1280}{2} = 640$m [1]

2. a) The back of the steel block [1]
   b) 90mm [1]
   c) It is smaller / half the size [1]

Page 160 The Electromagnetic Spectrum

1. a) Microwaves [1]
   b) Accept any sensible answer, e.g. X-rays for photographing bones OR gamma rays for sterilisation OR UV for sunbeds [2] (1 mark for wave, 1 mark for use)
2. frequency [1]; wavelength [1]
3. a) X-rays [1]
   b) It can penetrate soft tissue [1]; but is blocked by bone [1]

Page 161 Lenses

1. a) magnification = $\dfrac{2}{8}$ [1]; = 0.25 [1]
   b) magnification = $\dfrac{300}{2}$ [1]; = 150 [1]

   > Make sure both measurements are in the same units before calculating magnification.

2. Parallel rays of light entering the lens [1]; spread out / diverge when they leave the lens [1]
3. a) A convex / converging lens [1]
   b) More distant from the lens than the object [1]
4. a) Convex [1]
   b) Three correctly drawn ray lines with arrows [3] (1 mark for each line); correctly drawn image [1]

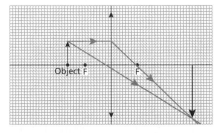

Object F   F

   c) magnification = $\dfrac{\text{height of image}}{\text{height of object}}$ = $\dfrac{20}{10}$ [1]; = 2 [1] (Allow correct calculation with readings from own diagram if drawn incorrectly)

Page 162 Light and Black Body Radiation

1. a) The red filter only lets red light through [1]; so only red light reaches the green filter, which only lets green light through, so no light gets through [1]
   b) A green circle [1]
2. a) That only infrared light is emitted / its wavelength it too long to be in the visible spectrum [1]
   b) It is cooler than the Sun [1]
   c) Radiation emitted increases in frequency as temperature increases [1]; only low frequency infrared is emitted [1]; so must be a lower temperature than the Sun, which emits higher frequency visible light [1]

Page 163 An Introduction to Electricity

1. current [1]; charge [1]; greater [1]; current [1]
2. a) potential difference = current × resistance / $V = IR$ [1]
   b) $230 = 5 \times R$ [1]; $R = \dfrac{230}{5} = 46\Omega$ [1]
   c) charge = current × time / $Q = It$ [1]
   d) $Q = 5 \times 120$ [1]; = 600C [1]
3. Energy [1]; greater [1]; current [1]
4. a) Closed switch [1]
   b) Battery [1]
   c) Fuse [1]
   d) Light dependent resistor / LDR [1]

Page 164 Circuits and Resistance

1. a) To adjust the resistance of the circuit [1]; and change the voltage across the component [1]
   b) Series [1]
   c) Parallel [1]
2. Four correctly drawn lines [3] (2 marks for two correct lines; 1 mark for one correct line)
   Light dependent resistor (LDR) – Resistance decreases as light intensity increases.
   Thermistor – Resistance decreases as temperature increases.
   Diode – Has a very high resistance in one direction.
   Filament light – Resistance increases as temperature increases.

Page 165 Circuits and Power

1. a) energy transferred = power × time / $E = Pt$ [1]; $E = 1600 \times 120$ [1]; = 192 000J [1]
   b) $\dfrac{192\,000}{100} \times 10$ [1]; = 19 200J [1] (accept any equivalent method)
2. a) $V = IR$, $V = 2 \times 3$ [1]; = 6V [1]
   b) $18 = 6 + V$ [1]; $V = 18 - 6 = 12$V [1]

   > The total potential difference in a series circuit is shared across the components.

   c) $18 = 2 \times R$ [1]; $R = \dfrac{18}{2} = 9\Omega$ [1]

Page 166 Domestic Uses of Electricity

1. a) 1.5V [1]; d.c. [1]
   b) 230V [1]; a.c. [1]
2. When a device is switched off, the live wire before the switch can still be at a non-zero potential [1]; touching this would create a potential difference between the wire and the ground [1];

Answers

this would make current flow through the person [1]; which would cause an electric shock [1]

Page 167 Electrical Energy in Devices

1. a) Kinetic [1]
   b) It is dissipated / lost [1]; to the surroundings [1]
2. a) Energy input = electrical [1]; useful energy output = kinetic [1]
   b) It disappears. [1]
   c) 500J [1]
3. The output from the generators goes through a step-up transformer [1]; this increases the voltage and also reduces the current [1]; the low current stops the cables from becoming hot [1]; which means less energy is lost during transmission [1]

Page 168 Static Electricity

1. a) True [1]
   b) False [1]
   c) False [1]
   d) True [1]
2. a) Friction [1]; has caused electrons to be removed from the rod / transferred to the jumper [1]; leaving the rod positively charged [1]
   b) repel [1]; attract [1]
3. charged [1]; electric [1]; force [1]; force [1]; charges [1]

Page 169 Magnetism and Electromagnetism

1. If free to move, the magnet will rotate so that the north pole of the magnet [1]; points to the Earth's north pole [1]
2. a) They will repel [1]
   b) It will be attracted to the magnet [1]
3. When the switch is pressed current flows in the electromagnet [1]; this magnetises the magnet [1]; which attracts the armature, causing the hammer to hit the gong [1]; the movement of the armature breaks the circuit, switching off the magnet [1]; the armature springs back and remakes the circuit, which starts the cycle again [1]

Page 170 The Motor Effect

1. The alternating electrical signal passes through the coil between the poles of the magnet [1]; which creates an alternating force on the coil [1]; the coil is fixed to the cone, so there is an alternating forces on the cone [1]; which vibrates at the same frequency as the alternating signal [1]

2. a) force on a conductor (at right-angles to a magnetic field carrying a current = magnetic flux density × current × length / $F = BIl$, $F = (0.3 \times 10^{-3}) \times 2 \times 0.05$ [1]; = 0.00 003N [1]
   b) There will be no force on the wire [1]

Page 171 Induced Potential and Transformers

1. a) A dynamo [1]
   b) The electricity produced is not constant [1] so the light brightness varies/gets dim at low speed/ goes off when stopped [1]
2. a) The amplitude of the waves would be twice as high [1]; and the frequency of the waves would be doubled [1]; so the period would be halved [1]
   b) The amplitude of the waves would be twice as high [1]; and the frequency of waves would be unchanged [1]
   c) The induced current will oppose the motion [1]; so the force needed to keep the alternator turning increases [1]

Page 172 Particle Model of Matter

1. a) The energy required [1]; to change 1kg of a substance from a solid to a liquid [1]
   b) thermal energy for a change of state = mass × specific latent heat / $E = mL$, $E = 0.012 \times (2.3 \times 10^6)$ [1]; = 27 600J [1]
2. a) The substance is condensing [1]; from a gas to a liquid [1]
   b) The particles are slowing down [1]; and the substance is cooling [1]

Page 173 Atoms and Isotopes

1. a) Electron, –1 [1]; neutron, 0 [1]; proton, +1 [1]
   b) It has the same number of protons [1]; as electrons [1]
   c) ion [1]; positive [1]
2. Path A is a long way from the nucleus and the alpha particle goes straight through [1]; Path B is close to the positive nucleus so the alpha particle is deflected [1]; Path C comes very close to the nucleus and the alpha particle is repelled back the way it came [1]

Two positively charged particles will repel each other.

Page 174 Nuclear Radiation

1. a) An unstable atom that gives out radiation [1]
   b) Beta decay [1]

c) It is stable / non-radioactive [1]
   d) Sodium [1]
2. Becquerel [1]
3. Gamma, beta, alpha [1]

Page 175 Using Radioactive Sources

1. a) No longer a risk [1]; because it has a half-life of just 8 days [1]; so would have completely decayed to the same level as background radiation in the last 30 years [1]
   b) Iodine-131 has a short half-life [1]; so it doesn't remain radioactive in the body for long [1]; Iodine-131 gathers in the thyroid so stays close to the cancer [1]; and beta radiation then destroys the cancer cells, as it is moderately ionising [1]
   c) 24 days is 3 half-lives [1]; 256 → 128 → 64 → 32, so a count rate of 32 remains [1]
2. It travels into the intestines and cannot pass the blockage [1]; a detector outside the body can detect the radiation [1]; no radiation will be detected after the blockage, so the location of the blockage can be found [1]

Page 176 Fission and Fusion

1. a) **Any two of:** rocks/soil [1]; radon gas [1]; food [1]; cosmic rays [1]
   b) The nuclear industry contributes a very small percentage of the total background radiation – much less than natural sources [1] (Accept any other sensible answer)
2. a) Alpha decay:
   $$^{219}_{86}\mathrm{Ra} \rightarrow {}^{215}_{84}\mathrm{Po} + {}^{4}_{2}\mathrm{He}\ [2]$$
   (1 mark for the correct radioactive particle on the left-hand side of the equation; 1 mark for the correct products)
   b) Beta decay:
   $$^{234}_{90}\mathrm{Th} \rightarrow {}^{234}_{91}\mathrm{Pa} + {}^{0}_{-1}\mathrm{e}\ [2]$$
   (1 mark for the correct radioactive particle on the left-hand side of the equation; 1 mark for the correct products)

Page 177 Stars and the Solar System

1. a) Gravity [1]
   b) The explosive force of fusion / radiation force [1]; and the compressive force of gravity [1]
   c) It expands [1]; to become a red giant [1]
   d) It could be engulfed by the star [1]
   e) They explode as a supernova and become a black hole or neutron star [1]
2. a) (1.007 825 × 4) – 4.0037 [1]; = 0.0276, so the relative mass falls by 0.0276 [1]

b) Eventually the decrease in mass will be significant enough **[1]**; that the gravitational attraction due to the Sun's mass will decrease, allowing the material to expand (as a red giant) **[1]**

Page 178 Orbital Motion and Red-Shift

1. a) The gravitational attraction of the Earth **[1]**
   b) Because the satellite is in circular motion **[1]**; the acceleration caused by the force makes the direction change, but the magnitude (speed) is unchanged **[1]**
   c) The radius of the orbit would change **[1]**
   d) At right-angles to the force of gravity **[1]**
2. a) The universe started with an explosion from a single point **[1]**; and has been expanding ever since **[1]**
   b) The light from a light source moving away becomes redder **[1]**; as its wavelength is stretched and becomes longer **[1]**
   c) All of the distant galaxies are moving away from one another **[1]**; and they appear to be moving away from the same place **[1]**

Pages 179–196 Practice Exam Paper 1

01.1 From left to right: 65J **[1]**; 5J **[1]**; 30J **[1]**
01.2 Heat the schools **[1]**; because it saves the most energy **[1]**; half of 5J is 2.5J **[1]**; but a quarter of 65J is over 15J **[1]**
01.3 efficiency =

$$\frac{\text{useful output energy transfer}}{\text{total input energy transfer}} \times 100\%,$$

$$= \frac{30}{100} \times 100\% = 30\% \quad \textbf{[1]}$$

02.1 It spreads out / is dissipated **[1]**; into the surroundings **[1]**
02.2 20 + 13 + 7 = 40% **[1]**
02.3 60% **[1]**
03.1 work done = force × distance / $W = Fs$, $W = 6 \times 2$ **[1]**; = 12J **[1]**
03.2 weight = mass × gravitational field strength / $W = mg$, $m = \frac{6}{10} = 0.6$kg **[1]**; gravitational potential energy = mass × gravitational field strength × height / $E_p = mgh$, $E_p = 0.6 \times 10 \times 2 = 12$J **[1]**
03.3 kinetic energy = 0.5 × mass × (speed)² / $E_k = \frac{1}{2}mv^2$ **[1]**; $v^2 = \frac{12}{0.5 \times 0.6} = 40$ **[1]**; $v = \sqrt{40} = 6.3$m/s **[1]**
04.1 Material A as it has the lowest density **[1]**; therefore it should be the best insulator as it contains more air spaces / will conduct the heat slowest **[1]**.

04.2 **Here is a sample answer worth 6 marks:** Wrap a test tube or beaker in material A. Fill the test tube or beaker with hot water at a certain temperature (e.g. 60°C). Start a stop watch and record the temperature of the water every minute until the temperature stops decreasing (reaches room temperature). Record the time taken for the water to reach room temperature. Repeat the investigation two more times and use the repeat results to calculate a mean time taken. Repeat the whole investigation for materials B and C, including taking repeat readings. Two variables that need to be controlled in this investigation are the thickness of the material and the volume of water used.
04.3 The insulation material may not totally cover the beaker or test tube, so heat may be lost through holes or gaps **[1]**; control this by ensuring the insulating material completely covers the beaker or test tube with only the thermometer poking out **[1]**.
   Or Time interval used may not be enough to determine an accurate end point **[1]**; measure the temperature at more regular intervals or use a temperature probe to read the temperature constantly **[1]**.
   Or Any other valid inaccuracy and improvement.
04.4 Hot water may causing scalding if spilled **[1]**; take care when pouring hot water into test tubes and beakers / ensure the water has cooled completely before moving the test tubes or beakers **[1]**.
05.1 The current starts at 1.5A **[1]**; decreases for 4ms **[1]**; then stabilises at 0.5A **[1]**
05.2 The resistance starts low at 8 ohms **[1]**; it increases rapidly **[1]**; until it reaches 24 ohms **[1]**
06.1 Y = variable resistor; Z = voltmeter **[2]**
06.2 To control the voltage applied to component X / adjust the resistance **[1]**
06.3 The voltage **[1]**
06.4 The current **[1]**
06.5 Accurately plotted voltage **[1]**; and current **[1]**; with all points joined by a smooth curve **[1]**

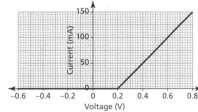

06.6 Yes, they are correct as all the points fit the line **[1]**

An anomalous result would be significantly higher or lower than the other results or would not fit the pattern.

06.7 A diode **[1]**
07.1 When rubbed they became charged **[1]**; they are made of different materials and have opposite charges, which attract **[1]**
07.2 They still attract **[1]**
07.3 They will repel **[1]**
07.4 They are made of the same material, so end up with the same charge **[1]**; and like charges repel **[1]**
07.5 When an insulating material is rubbed, electrons move **[1]**; either from the rod to the cloth, leaving the rod positive **[1]**; or from the cloth to the rod, making the rod negative **[1]**
08.1 2 + 1 + 4 = 7A **[1]**
08.2 $P = 230 \times 7$ **[1]**; = 1610 **[1]**
08.3 This could damage / overload the circuit **[1]**; and cause an electric shock **[1]**
08.4 power = (current)² × resistance / $P = I^2R$, 1610 = 72 × R **[1]**; $R = \frac{1610}{49} = 32.9\Omega$ **[1]**
08.5 charge flow = current × time / $Q = It$, $Q = 7 \times 60 \times 5$ **[1]**; = 2100C **[1]**
09.1 They have the same number of protons **[1]**
09.2 A large unstable nucleus **[1]**; splits into two or more smaller nuclei and releases energy **[1]**
09.3 Beta decay **[1]**
09.4 The mass number has not changed during emission, but the proton number has increased by one **[1]**
09.5 The time it takes for half of the radioactive isotopes to decay / for the count rate to halve **[1]**
09.6 $\frac{1}{8}$ means that 3 half-lives have passed **[1]**; so is (30 × 3 =) 90 years **[1]**
09.7 It is more hazardous because it decays quickly / is highly active, so gives out radiation quickly **[1]**; but is not radioactive for a long time, so in the long-term it is less hazardous **[1]**
10.1 kinetic energy = 0.5 × mass × (speed)² / $E_k = \frac{1}{2}mv^2$, $m = \frac{E_k}{\frac{1}{2}v^2}$ **[1]**; $m = \frac{472500}{450}$ **[1]**; = 1050 **[1]**; final answer = 1050kg (unit must be correctly stated for mark) **[1]**
10.2 The bus has more mass **[1]**; so it has more kinetic energy **[1]**
10.3 kinetic energy = 0.5 × mass × (speed)² / $E_k = \frac{1}{2}mv^2$ **[1]**; increase in E_k = 0.5 × 1200 × (14² – 8²) **[1]**; = 79 200J **[1]**
10.4 Because the kinetic energy depends on the square of the speed **[1]**

Answers

11.1 gravitational potential energy = mass × gravitational field strength × height / $E_p = mgh$, $E_p = 65 \times 10 \times 1.25$ **[1]**; = 812.5J **[1]**

11.2 kinetic energy = 0.5 × mass × (speed)^2 / $E_k = \frac{1}{2}mv^2$, $812.5 = 0.5 \times 65 \times v^2$ **[1]**; $v^2 = 25$ **[1]**; $v = \sqrt{25}$ = 5m/s **[1]**

12.1 12V **[1]**

12.2 2 + 2 = 4A **[1]**

12.3 5A **[1]**

12.4 potential difference = current × resistance / $V = IR$, $R = \frac{12}{2}$ **[1]**; = 6Ω **[1]**

12.5 It is less / half **[1]**

12.6 power = potential difference × current / $P = VI$, $P = 12 \times 2$ **[1]**; = 24W each **[1]**

12.7 $\frac{96\,000}{(24 \times 2)}$ **[1]**; = 2000s **[1]**

Pages 197–214 Practice Exam Paper 2

01.1 So it will be attracted by the magnetic coil **[1]**

01.2 It will need to increase **[1]**

01.3 A bigger mass makes a bigger force pulling down on the left, so a bigger force is needed on the right **[1]**; a bigger current will increase the strength of the magnetic field created by the coil **[1]**

01.4 It will increase the mass that can be balanced **[1]**; because the iron core will make the magnet stronger **[1]**

01.5 Moving the mass closer to the pivot **[1]**; Increasing the number of batteries **[1]**

01.6 Moment of force = force × distance / $M = Fd$, $M = 50 \times 0.05$ **[1]**; = 2.5Nm **[1]**

01.7 $2.5 = F \times 0.2$ **[1]**; $F = \frac{2.5}{0.2}$ = 12.5N **[1]**

02.1 Towards the centre of the circle **[1]**

02.2 The tension of the string **[1]**

02.3 The radius of the circle **[1]**; and the mass of the bung **[1]**

02.4 speed; velocity; direction **[1]**

02.5 Gravity / the gravitational attraction of the Earth on the Moon **[1]**

03.1 Missing voltmeter reading = 1.6V **[1]**; missing potential difference of supply = 12.8V **[1]**

03.2 An alternating potential difference **[1]**; across the primary coil **[1]**; induces an alternating magnetic field in the core **[1]**; this alternating field induces an alternating potential difference in the secondary coil **[1]**

04.1 pressure = $\frac{\text{force normal to the surface}}{\text{area of that surface}}$ / $P = \frac{F}{A}$ **[1]**; $P = \frac{20}{0.006}$ = 3333.33 **[1]**; = 3300N/m^2 **[1]**

04.2 Hole C **[1]**

04.3 It has the least weight of water above it **[1]**; so is the lowest pressure **[1]**

05 Turning the key switches the electromagnet on **[1]**; this attracts the pivoted armature **[1]**; which pushes the contacts together, completing the starter motor circuit **[1]**

06.1 Gravity pulls clouds of hydrogen gas together **[1]**; once there is enough mass the star becomes hot and dense enough for fusion to start **[1]**

06.2 The explosive / expansive forces of fusion are balanced **[1]**; by the compressive / attractive force of gravity **[1]**

06.3 The bigger the star the faster the rate of fusion **[1]**

06.4 The Big Bang theory **[1]**

06.5 In stars, larger elements are formed during fusion **[1]**; the largest stars can fuse bigger elements and the heaviest elements are formed during a supernova **[1]**

07.1 As the magnet moves into the coil **[1]**; the magnetic field lines are cut by the coil **[1]**; this induces a potential difference in the coil **[1]**; and because the coil is part of a complete circuit a current flows, which is measured on the ammeter **[1]**

07.2 **From top to bottom:** ammeter reads zero **[1]**; ammeter reading is high **[1]**; ammeter reading is low in the opposite direction to the original situation **[1]**

07.3 A north pole **[1]**

08.1 The output voltage is lower than the input voltage **[1]**

08.2 The transformer is made from an iron core **[1]**; the input voltage is supplied to the primary coil **[1]**; and the output voltage induced in the secondary coil **[1]**; there are more turns on the primary coil than on the secondary coil **[1]**

08.3 $\frac{\text{potential difference across primary coil}}{\text{potential difference across secondary coil}}$ $= \frac{\text{number of turns in primary coil}}{\text{number of turns in secondary coil}}$, $\frac{230}{4.6} = \frac{2000}{\text{number of turns in secondary coil}}$ **[1]**; number of turns in secondary coil $= \frac{2000}{50}$ = 40 **[1]**

09.1 20Hz **[1]**; 20 000Hz **[1]**

09.2 Sound with a frequency greater than 20 000Hz **[1]**

09.3 wave speed = frequency × wavelength / $v = f\lambda$, $\lambda = \frac{1500}{2\,000\,000}$ **[1]**; = 0.00075m **[1]**

09.4 The ultrasound is transmitted into the body and is reflected back if kidney stones are present **[1]**; the time delay before the reflected wave is detected by the receiver indicates where the stones are **[1]**

10.1 magnified **[1]**; upright **[1]**

10.2 3 **[1]**

11.1 The amplitude of a wave is the maximum displacement of a point on a wave away from its undisturbed position **[1]**; the wavelength of a wave is the distance from a point on one wave to the equivalent point on the adjacent wave **[1]**.

11.2 **Here is a sample answer worth 6 marks:** Produce a wave in the ripple tank. Time how long it takes this wave to travel the length of the tank. Use this time to calculate the wave speed using the formula speed = distance ÷ time. To find the frequency, count the number of waves which pass a fixed point in a given time (e.g. 10 seconds). Divide this number by the time to give the frequency. Use a ruler to estimate the distance between the peaks of the wave as it travels. Repeat the experiment several times and then use the repeat readings to calculate mean values for the frequency, wavelength and speed. Sources of inaccuracy include it being difficult to count and measure small or fast waves. Estimating the distance with a ruler is also an inaccurate way to determine wavelength. A stroboscope could be used in the investigation to improve accuracy.

11.3 The water depth will affect the wave speed and wavelength at a given frequency **[1]**; so changes in depth along the tank will lead to changes in the speed and wavelength as the wave moves along the tank **[1]**.

11.4 Water could splash out of the tank causing a potential slip hazard **[1]**; control this by ensuring that, when waves are produced, water does not splash out and any spills are wiped up immediately **[1]**

Or Using a stroboscope presents a risk to people with photosensitive epilepsy **[1]**; to control this, ensure that no one at risk is in the room when the stroboscope is used **[1]**.

Or **Any other valid hazard and control.**

12 A correctly drawn diagram showing three accurately drawn ray lines **[3]**; and a correctly drawn arrow to represent the image **[1]**

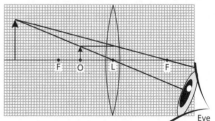

13.1 **From top to bottom:** 1500m, 3×10^{-8}m, 5×10^{-7}m, 1×10^{-11}m **[3]** (2 marks for two in the correct position; 1 mark for one correct)

13.2 wave speed = frequency × wavelength / $v = f\lambda$ **[1]**; $f = \dfrac{3 \times 10^8}{300}$ **[1]**; $= 1 \times 10^6$Hz **[1]**

13.3 acceleration = $\dfrac{\text{change in velocity}}{\text{time}}$ / $a = \dfrac{\Delta v}{t}$ **[1]**; $a = \dfrac{10}{4}$ **[1]**; $= 2.5$m/s^2 **[1]**

13.4 10m/s **[1]**

13.5 distance travelled = speed × time / $s = vt$ **[1]**; $s = 10 \times 4 = 40$m **[1]**

13.6 resultant force = mass × acceleration / $F = ma$, $F = 1.5 \times 25$ **[1]**; $= 37.5$N **[1]**

14.1 force = $\dfrac{\text{change in momentum}}{\text{time taken}}$ / $F = \dfrac{m\Delta v}{t}$ **[1]**; change in momentum = $4000 \times 0.004 = 16$kg m/s **[1]**

14.2 momentum = mass × velocity / $p = mv$ **[1]**; $16 = 0.05 \times$ velocity, velocity = $\dfrac{16}{0.05} = 320$m/s **[1]**

Remember to change the mass from grams to kilograms before carrying out the calculation.

Notes

Notes

Notes

GCSE Physics Workbook